ENVIRONMENTAL HEALTH -
PHYSICAL, CHEMICAL AND BIOLOGICAL FACTORS

HEALTH ASSESSMENTS OF CHEMICALS AND CONTAMINANTS

ANALYSES OF THE EPA'S IRIS PROGRAM

ENVIRONMENTAL HEALTH - PHYSICAL, CHEMICAL AND BIOLOGICAL FACTORS

Additional books in this series can be found on Nova's website under the Series tab.

Additional e-books in this series can be found on Nova's website under the e-book tab.

ENVIRONMENTAL HEALTH -
PHYSICAL, CHEMICAL AND BIOLOGICAL FACTORS

HEALTH ASSESSMENTS OF CHEMICALS AND CONTAMINANTS

ANALYSES OF THE EPA'S IRIS PROGRAM

MARCUS ROWE
EDITOR

New York

For permission to use material from this book please contact us:
Telephone 631-231-7269; Fax 631-231-8175
Web Site: http://www.novapublishers.com

Library of Congress Cataloging-in-Publication Data

ISBN: 978-1-63321-166-7

Published by Nova Science Publishers, Inc. † New York

CONTENTS

PREFACE

The Environmental Protection Agency (EPA) has not conducted a recent evaluation of demand for Integrated Risk Information System (IRIS) toxicity assessments with input from users inside and outside EPA. The health effects information in IRIS provides fundamental scientific information EPA needs to develop human health risk assessments. Specifically, EPA issued a needs assessment report in 2003, which estimated that 50 new or updated IRIS toxicity assessments were needed each year to meet users' needs. This book determines the extent to which EPA has evaluated demand for IRIS toxicity assessments from users inside and outside EPA; EPA's process for nominating and selecting chemicals for IRIS toxicity assessment and strategies for addressing any unmet agency needs when IRIS toxicity assessments are not available, applicable, or current.

Chapter 1 – EPA created the IRIS database in 1985 to help develop consensus opinions within the agency about the health effects from chronic exposure to chemicals. The health effects information in IRIS—referred to as IRIS toxicity assessments—provides fundamental scientific information EPA needs to develop human health risk assessments. GAO was asked to review the effectiveness of EPA's implementation of its IRIS toxicity assessment process. This report determines the extent to which (1) EPA has evaluated demand for IRIS toxicity assessments from users inside and outside EPA; (2) EPA's process for nominating and selecting chemicals for IRIS toxicity assessment accurately reflects demand; and (3) EPA has implemented a strategy for addressing any unmet agency needs when IRIS toxicity assessments are not available, applicable, or current. To do this work, GAO reviewed and analyzed IRIS nomination data, among other things, and interviewed EPA officials. GAO did not evaluate the scientific content or quality of IRIS toxicity assessments.

Chapter 2 – The U.S. Environmental Protection Agency's (EPA) Integrated Risk Information System (IRIS) Program provides health effects information on chemicals to which the public may be exposed, providing a critical part of the scientific foundation for EPA's decisions to protect public health. EPA has made several changes to this important Program over the past few years, streamlining the assessment development process, improving transparency, and creating efficiencies within the Program. In April 2011, the National Research Council (NRC) made several recommendations to EPA for improving the development of IRIS assessments. The NRC's recommendations were focused on the development of draft assessments, and the NRC was clear that their intent was not to delay assessments.

EPA has made progress in implementing these recommendations. Consistent with the advice of the NRC, EPA is implementing these recommendations using a phased approach and is making the most extensive changes to documents that are in the earlier steps of the assessment development process. For assessments that are in the later stages of development, including assessments that have been posted on the IRIS database since the release of the NRC report, EPA is implementing the recommendations as feasible without taking the assessments backwards to earlier steps of the process. Phase 1 of implementing the NRC recommendations has focused on editing and streamlining documents and using more tables, figures, and appendices. EPA is now in Phase 2 of implementing the NRC recommendations and will soon publicly release two draft IRIS assessments that represent a major advancement in implementing the NRC recommendations. EPA is using a new document structure for these draft assessments, including an Executive Summary presenting major conclusions, a Preamble describing methods used to develop the assessment, distinct sections on Hazard Identification and Dose-Response Analysis, and more tables and figures to clearly present data. Additionally as part of Phase **2,** Phase 2, EPA is addressing all of the short-term recommendations provided by the NRC. As part of this effort, EPA will make several changes to IRIS assessments. Highlights include: evaluating and describing the strengths and weaknesses of critical studies in a more uniform way; including toxicity values for multiple effects associated with the chemical, if applicable and where the data allow; routinely considering the use of multiple data sets of combined multiple responses in deriving toxicity values, where appropriate; and evaluating existing guidelines to establish clearer criteria for study selection.

Phase 3 of implementation will incorporate the longer-term scientific recommendations made by the NRC.

The U.S. Congress has directed EPA to issue a progress report to the House and Senate Committees on Appropriations and relevant Congressional authorizing committees to describe EPA's implementation Of the NRC recommendations. This report provides Congress, stakeholders, and the public with an update on the IRIS Program and EPA's progress toward implementing the NRC recommendations and improving the Program.

Chapter 3 – This report summarizes the Toxic Substances Control Act (TSCA) and the major regulatory programs dealing with chemical production and distribution in U.S. commerce.

The President's Council on Environmental Quality proposed comprehensive federal legislation in 1971 to identify and control potentially dangerous chemicals in U.S. commerce that were not adequately regulated under other environmental statutes. President Ford signed TSCA into law on October 11, 1976. Subsequently, five titles were added to address specific concerns—asbestos in 1986 (Title II, P.L. 99-519), radon in 1988 (Title III, P.L. 100-551), lead in 1992 (Title IV, P.L. 102-550), environmental and energy issues in schools in 2007 (Title V, P.L. 110-140), and formaldehyde emissions from composite wood products in 2010 (Title VI, P.L. 111-199). In 2008, Congress added provisions to Title I, Section 6 and Section 12, banning certain activities with respect to elemental mercury (P.L. 110-414).

TSCA authorizes EPA to identify potentially dangerous chemicals in U.S. commerce that should be subject to federal control. The act authorizes EPA to gather and disseminate information about production, use, and possible adverse effects to human health and the environment of existing chemicals, and to issue "test rules" that require manufacturers and processors of potentially dangerous chemicals to conduct and report the results of scientific studies to fill information gaps. For chemicals new to U.S. commerce, TSCA requires pre-market screening and regulatory tracking of new chemical products.

If EPA identifies unreasonable risks associated with existing or new chemicals, TSCA requires the agency to initiate rulemaking to reduce risks to a reasonable level. EPA may regulate the manufacture, importation, processing, distribution, use, and/or disposal of chemicals. TSCA provides a variety of regulatory tools to EPA, ranging in severity from a total ban on production, import, and use to a requirement that a product must bear a warning label at the point of sale. However, TSCA directs EPA to use the least burdensome option that can reduce risk to a level that is reasonable, given the benefits provided by the chemical product or process.

Title I of the original statute establishes the core program, directs EPA to control risks from polychlorinated biphenyls (PCBs), and bans certain activities with respect to elemental mercury. Title II directs EPA to set standards for asbestos mitigation in schools and requires asbestos contractors to be trained and certified. Title III directs EPA to provide technical assistance to states that choose to support radon monitoring and control. Title IV provides similar assistance with respect to abatement of lead-based paint hazards. Title V addresses environmental issues at schools, including energy efficiency. Finally, Title VI directs EPA to set standards for emissions of formaldehyde from composite wood products.

Chapter 4 – EPA's Integrated Risk Information System (IRIS) is a human health assessment program that evaluates risk information on effects that may result from exposure to environmental contaminants. Through the IRIS Program, EPA provides the highest quality science-based human health assessments to support the Agency's regulatory activities. The IRIS database contains information for more than 540 chemical substances containing information on human health effects that may result from exposure to various substances in the environment. IRIS is prepared and maintained by the EPA's National Center for Environmental Assessment (NCEA) within the Office of Research and Development (ORD).

The heart of the IRIS system is its collection of searchable documents that describe the health effects of individual substances and that contain descriptive and quantitative information in the following categories:

- *Noncancer effects:* Oral reference doses and inhalation reference concentrations (RfDs and RfCs (http://www.epa.gov/iris/help_ gloss. htm#r), respectively) for effects known or assumed to be produced through a nonlinear (possibly threshold) mode of action. In most instances, RfDs and RfCs are developed for the noncarcinogenic effects of substances.
- *Cancer effects:* Descriptors that characterize the weight of evidence for human carcinogenicity, oral slope factors, and oral and inhalation unit risks for carcinogenic effects. Where a nonlinear mode of action is established, RfD and RfC values may be used.

In: Health Assessments …
Editor: Marcus Rowe

ISBN: 978-1-63321-166-7

Chapter 1

CHEMICAL ASSESSMENTS: AN AGENCYWIDE STRATEGY MAY HELP EPA ADDRESS UNMET NEEDS FOR INTEGRATED RISK INFORMATION SYSTEM ASSESSMENTS*

United States Government Accountability Office

WHY GAO DID THIS STUDY

EPA created the IRIS database in 1985 to help develop consensus opinions within the agency about the health effects from chronic exposure to chemicals. The health effects information in IRIS—referred to as IRIS toxicity assessments—provides fundamental scientific information EPA needs to develop human health risk assessments. GAO was asked to review the effectiveness of EPA's implementation of its IRIS toxicity assessment process. This report determines the extent to which (1) EPA has evaluated demand for IRIS toxicity assessments from users inside and outside EPA; (2) EPA's process for nominating and selecting chemicals for IRIS toxicity assessment accurately reflects demand; and (3) EPA has implemented a strategy for addressing any unmet agency needs when IRIS toxicity assessments are not

* This is an edited, reformatted and augmented version of United States Government Accountability Office Publication, No. GAO-13-369, dated May 2013.

available, applicable, or current. To do this work, GAO reviewed and analyzed IRIS nomination data, among other things, and interviewed EPA officials. GAO did not evaluate the scientific content or quality of IRIS toxicity assessments.

WHAT GAO RECOMMENDS

GAO recommends that EPA evaluate demand for IRIS assessments; document how the agency applies its selection criteria, including the circumstances under which an IRIS toxicity assessment is or is not needed and; develop an agencywide strategy including, at a minimum, coordination across EPA offices, as well as with other federal agencies, to identify and fill data gaps, and providing guidance that describes alternative sources of toxicity information. EPA agreed with the first two recommendations and partially agreed with the third.

WHAT GAO FOUND

The Environmental Protection Agency (EPA) has not conducted a recent evaluation of demand for Integrated Risk Information System (IRIS) toxicity assessments with input from users inside and outside EPA. Specifically, EPA issued a needs assessment report in 2003, which estimated that 50 new or updated IRIS toxicity assessments were needed each year to meet users' needs. However, GAO did not find sufficient support for the estimate. In addition, IRIS Program officials recognize that the 2003 estimate does not reflect current conditions, but the agency does not plan to perform another evaluation of demand. Without a clear understanding of current demand for IRIS toxicity assessments, EPA cannot adequately measure the program's performance; effectively determine the number of IRIS toxicity assessments required to meet the needs of IRIS users; or know the extent of unmet demand.

The IRIS Program's chemical nomination and selection process, which the agency uses to gauge interest in the IRIS Program from users inside and outside of EPA, may not accurately reflect current demand for IRIS toxicity assessments. The 75 chemicals that were nominated in response to EPA's most recent 2011 nomination period may not reflect demand for a number of reasons. For example, given the long-standing challenges the IRIS Program has had in routinely starting new assessments, according to some EPA IRIS users, they chose not to nominate new chemicals for assessment. Also, EPA has not clearly articulated how the IRIS Program applies the criteria it uses to prioritize the selection of chemicals for IRIS toxicity assessment—including how it determines the circumstances under which an IRIS toxicity assessment is or is not needed. Consequently, for chemicals that were nominated but not selected for assessment, it is not clear how many, if any, were excluded from consideration because they did not meet the IRIS Program's selection criteria because the IRIS Program determined that an IRIS toxicity assessment was not needed—or, alternatively, if they were not selected due to resource constraints or other reasons.

EPA has not implemented an agencywide strategy for addressing the unmet needs of EPA program offices and regions when IRIS toxicity assessments are not available, applicable, or current. Specifically, EPA does not have a strategy for identifying and filling data gaps that would enable it to conduct IRIS toxicity assessments for nominated chemicals that are not selected for assessment because sufficient data from health studies are not available. IRIS Program officials stated that no agencywide mechanism exists for EPA to ensure that chemicals without sufficient scientific data during one nomination period will have such information by the next nomination period or even the one after that. These officials acknowledged that better coordination across EPA and with other federal agencies could help address the issue. EPA also does not have agencywide guidance for addressing unmet needs when IRIS toxicity assessments are not available, applicable, or current. In the absence of agencywide guidance, officials from select EPA offices stated that they used a variety of alternatives to IRIS toxicity assessments to meet their needs, including using toxicity information from other EPA offices or other federal agencies.

ABBREVIATIONS

ATSDR	Agency for Toxic Substances and Disease Registry
Cal/EPA	California Environmental Protection Agency
CAS	Chemical Abstracts Service
DIPE	diisopropyl ether
EPA	Environmental Protection Agency
ETBE	ethyl tertiary butyl ether
GPRA	Government Performance and Results Act of 1993

IRIS	Integrated Risk Information System
MTBE	methyl tertiary butyl ether
OMB	Office of Management and Budget
PART	Program Assessment Rating Tool
PPRTV	Provisional Peer Reviewed Toxicity Value
TAME	tertiary amyl methyl ether
TBA	tertiary butyl alcohol

May 10, 2013

The Honorable Barbara Boxer
Chairman
Committee on Environment and Public Works
United States Senate

Dear Madam Chairman:

The Environmental Protection Agency's (EPA) Integrated Risk Information System (IRIS)—a database integral to the agency's mission of protecting human health and the environment—contains EPA's scientific position on the potential human health effects that may result from exposure to various chemicals in the environment.[1] EPA created the IRIS database in 1985 to help it develop consensus opinions within the agency about the health effects from chronic exposure to chemicals and currently includes information on more than 550 chemicals. The health effects information in the IRIS database—referred to as IRIS toxicity assessments—provides fundamental scientific information EPA needs to develop human health risk assessments on particular chemicals.[2] These human health risk assessments, in turn, provide the foundation for EPA's risk management decisions, such as whether EPA should establish air and water quality standards to protect the public from exposure to particular toxic chemicals. EPA's IRIS Program develops new IRIS toxicity assessments and, as needed, updates information in existing IRIS toxicity assessments contained in the IRIS database.[3] The importance of EPA's IRIS Program has increased over time as EPA program offices and regions have increasingly relied on IRIS toxicity assessments in making environmental protection and risk management decisions. In addition, state and local environmental programs, as well as some international regulatory bodies, rely on IRIS toxicity assessments in managing their environmental protection programs.

Although the information in the IRIS database is a critical primary component of EPA's capacity to support scientifically sound decisions, policies, and regulations, we have reported previously on EPA's difficulty producing timely, credible IRIS toxicity assessments. Specifically, in March 2008, we reported that the IRIS database was at serious risk of becoming obsolete because EPA had not been able to (1) keep its existing assessments current or (2) complete assessments of the most important chemicals of concern.[4] Further, we reported that because EPA staff time continued to be dedicated to completing ongoing assessments, EPA's ability to both keep existing assessments (at that time more than 540) up to date and initiate new assessments was limited. We also reported that, although the number of program staff quadrupled from 8 to 37 and program funding increased from $1.7 million to $9.6 million for fiscal years 2000 to 2007, EPA had on average completed about five assessments annually during this period. For these and other reasons, in 2009 we added EPA's processes for assessing and controlling toxic chemicals to our list of areas at high risk for waste, fraud, abuse, and mismanagement or in need of broad-based transformation.[5] In response to our 2008 report and subsequent high-risk designation, EPA revised its IRIS toxicity assessment process in May 2009.

In December 2011, we reported, among other things, that EPA's initial productivity gains under the May 2009 revised process had not been sustained and that the agency continued to face challenges in implementing the IRIS Program. 6 For example, from May 2009 through September 2011, EPA completed 20 IRIS toxicity assessments—more than doubling the total productivity it achieved during the two fiscal years immediately preceding the adoption of the revised process. However, 16 of these toxicity assessments were completed in the first year and a half of implementing the revised process, and productivity fell sharply during fiscal year 2011, with EPA issuing 4 IRIS toxicity assessments. We reported that, even if EPA were to overcome the significant productivity difficulties it has experienced in recent years and met its goal of completing 40 assessments in fiscal year 2012, it was not clear that this level of productivity would meet the needs of IRIS users inside and outside EPA.[7] We noted that beyond EPA's ongoing assessments, and some that were on hold, the demand for additional IRIS toxicity assessments was unclear.

With tens of thousands of chemicals listed with EPA for commercial use in the United States, and about 1,000 new chemicals listed for commercial use each year, demand for IRIS toxicity assessments is potentially very high. In this context, you asked us to review the effectiveness of EPA's implementation of its IRIS toxicity assessment process. Our objectives were to determine the extent to which (1) EPA has evaluated demand for IRIS toxicity assessments from users inside and outside EPA; (2) EPA's process for nominating and selecting chemicals for IRIS toxicity assessment accurately reflects demand; and (3) EPA has implemented a strategy for addressing any unmet needs of EPA program offices and regions when IRIS toxicity assessments are not available, applicable, or current.

To determine the extent to which EPA has evaluated demand for IRIS toxicity assessments, we reviewed EPA's 2003 evaluation of demand for IRIS toxicity assessments. We also interviewed officials at the National Center for Environmental Assessment, which manages the IRIS Program and develops IRIS toxicity assessments (henceforth IRIS Program officials). Specifically, we interviewed these officials to determine whether they had conducted other evaluations of demand since 2003, how they derived the 2003 estimate, and whether that estimate reflects current conditions. We reviewed the report to determine its basis for estimating IRIS demand and the number of assessments it needed to conduct in order to meet demand.

To determine the extent to which EPA's process for nominating and selecting chemicals for IRIS toxicity assessments accurately reflects current demand, we reviewed data provided by the IRIS Program and from the IRIS Program's website on the number of IRIS toxicity assessments completed annually from fiscal years 2002 through 2012. In addition, we analyzed all chemical nomination forms submitted by EPA program offices and regions to the IRIS Program from 2005, 2007, and 2011—which were the last three times that the IRIS Program solicited nominations for new and updated IRIS toxicity assessments.[8] For additional perspective on user needs, we reviewed non-EPA IRIS users' chemical nomination forms from 2011. In addition, we reviewed the IRIS Program's processes for soliciting nominations and selecting chemicals for toxicity assessments, including the agency's selection criteria. We also reviewed agency guidance and interviewed IRIS Program officials to better understand the chemical nomination and selection process.

To determine the extent to which EPA has implemented a strategy for addressing any unmet needs of EPA program offices and regions when IRIS toxicity assessments are not available, applicable, or current, we reviewed the IRIS Program's efforts to analyze IRIS user chemical nominations. For context, we interviewed IRIS Program officials. For additional perspective, we interviewed officials using a standard set of questions from a nonprobability sample of three EPA program offices and one region: the Office of the Administrator, the Office of Water, the Office of Solid Waste and Emergency Response, and EPA's Region 2.[9] We selected these program offices and region because they submitted 78 percent of the chemical nominations to the IRIS Program during the period we reviewed—2005, 2007, and 2011. These offices and region ranked the highest in terms of the number of chemical nominations submitted and, in some cases, nominated chemicals more than once during different nomination years. Because this is a nonprobability sample, it is not generalizable to all EPA program offices and regions, but it can provide illustrative examples of the experience of those EPA program offices and one region that nominated 78 percent of chemical nominations for IRIS toxicity assessment during the period we reviewed. For example, we received information from officials from these offices about how EPA program offices and one region nominate chemicals for IRIS toxicity assessment, how the IRIS Program meets the needs of these offices and region over the course of nomination periods, and what alternative toxicity assessments these offices and region turn to when IRIS toxicity assessments are not available.

Separately from our nonprobability sample, we also interviewed officials from the Office of Pollution Prevention and Toxics, within the Office of Chemical Safety and Pollution Prevention, because it did not nominate any chemicals for IRIS toxicity assessment for any of the last three nomination periods. We did not evaluate the scientific content or quality of IRIS toxicity assessments. A more detailed description of our scope and methodology can be found in appendix I. For a summary of approaches used by selected EPA program offices and regions to address their unmet needs regarding IRIS toxicity assessments, see appendix II. We conducted this performance audit from April 2012 to May 2013 in accordance with generally accepted government auditing standards. Those standards require that we plan and perform the audit to obtain sufficient, appropriate evidence to provide a reasonable basis for our findings and conclusions based on our audit objectives. We believe that the evidence obtained provides a reasonable basis for our findings and conclusions based on our audit objectives.

BACKGROUND

This section discusses EPA's human health risk assessment and risk management practices and its processes for soliciting nominations and selecting chemicals for new and updated IRIS toxicity assessments.

EPA's Human Health Risk Assessment and Risk Management Practices

EPA's ability to effectively implement its mission of protecting public health and the environment is critically dependent on credible and timely assessments of the risks posed by chemicals. Such assessments are the cornerstone of scientifically sound environmental decisions, policies, and regulations under a variety of statutes, such as the Safe Drinking Water Act, the Toxic Substances Control Act, and the Clean Air Act. EPA assesses the human health risks of chemicals using a model from the National Academies.[10] This model includes four components: (1) hazard identification, (2) dose-response assessment, (3) exposure assessment, and (4) risk characterization (see fig. 1).

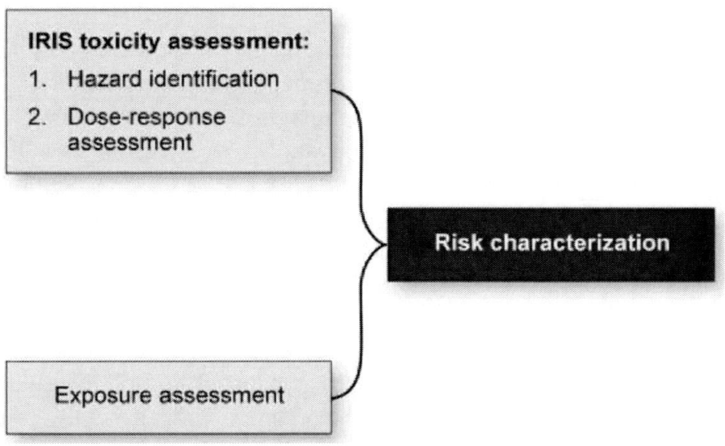

Source: GAO presentation of the risk assessment component of the National Academies' risk assessment and risk management model used by EPA.

Figure 1. National Academies' Risk Assessment Model Used by EPA.

For some, but not all chemicals, EPA conducts the first two sequential analyses of a human health risk assessment—that is, the hazard identification and dose-response assessment—under its IRIS Program. Taken together, these two steps are commonly referred to as IRIS toxicity assessments. EPA's IRIS Program—managed by EPA's National Center for Environmental Assessment within the Office of Research and Development—develops new IRIS toxicity assessments and updates existing IRIS toxicity assessments if revisions are warranted on the basis of newly published peer-reviewed studies.

EPA program offices and regions combine information from IRIS toxicity assessments with the results from chemical exposure assessments to characterize risk, which provides information on the probability that the adverse effects described in hazard identification will occur under the conditions described in the exposure assessment.[11] These four steps— hazard identification, dose-response assessment, exposure assessment, and risk characterization—comprise human health risk assessments, which EPA offices use to make risk management decisions. Risk management, as opposed to risk assessment, involves integrating the risk assessment information with other information—such as economic information on the costs and benefits of mitigating a risk, technological information on the feasibility of managing the risk, and the concerns of various stakeholders—to determine whether the health risks identified in a chemical risk assessment warrant EPA taking regulatory or other risk management actions.

A typical IRIS toxicity assessment contains a qualitative hazard identification and quantitative dose-response assessment. The qualitative hazard identification identifies noncancer and cancer health effects that may be caused by exposure to a given chemical. For cancer effects, EPA qualitatively describes the carcinogenic potential of a chemical in a narrative that includes selecting a weight-of-evidence descriptor, ranging from "carcinogenic to humans" to "not likely to be carcinogenic to humans."[12]

Following hazard identification, a dose-response assessment is conducted for both noncancer and cancer effects assuming adequate data are available. A quantitative dose-response assessment characterizes the quantitative relationship between the exposure to a chemical and the resultant health effects. The quantitative dose-response assessment relies on experimental data, primarily from either animal (toxicity) or human (epidemiology) studies. The noncancer dose-response assessment may include the following:

- an oral reference dose—an estimate (with uncertainty spanning perhaps an order of magnitude) of the daily oral exposure to a chemical that is likely to be without an appreciable risk of deleterious effects during a person's lifetime—expressed in terms of milligrams per kilogram per day, and
- an inhalation reference concentration—an estimate (with uncertainty spanning perhaps an order of magnitude) of the continuous inhalation exposure to a chemical that is likely to be without an appreciable risk

of deleterious effects during a person's lifetime—expressed in terms of milligrams per cubic meter.

According to EPA officials, the quantitative cancer dose-response assessment typically includes estimates of a chemical's carcinogenic potency by both the oral and inhalation routes of exposure. For oral exposures, the "oral slope factor" is an estimated 95 percent upper bound on the increased cancer risk per increased unit of exposure (in mg/kgday) to a chemical over a lifetime. For inhalation exposures, the "inhalation unit risk" is an estimated 95 percent upper bound on the increased cancer risk per increased unit of exposure (in $\mu g/m^3$ in air) to a chemical over a lifetime. The toxicity values derived in both noncancer and cancer dose-response assessments—that is, the oral reference dose, inhalation reference concentration, oral slope factor, and inhalation unit risk—are often referred to as IRIS values.

IRIS toxicity assessments estimate the potential health effects of lifelong (chronic) exposure to chemicals. According to the Office of Management and Budget (OMB), the IRIS Program is the only federal program that provides qualitative and quantitative assessments of both cancer risks and noncancer effects of chemicals.[13] In addition, according to EPA's Human Health Risk Assessment Strategic Research Action Plan, no other federal health assessment program has (1) a similar mission and scope or (2) internal and external peer review processes that are as rigorous.[14] Specifically, the IRIS toxicity assessment process includes internal EPA review; two interagency reviews by other federal agencies and White House offices (e.g., OMB); public review and comment; and a rigorous, independent, external peer review.[15] However, IRIS is not the only source of toxicity information available to EPA program offices and regions. For many chemicals, IRIS toxicity assessments are not available, applicable, or current; therefore, in some cases, EPA program offices and regions rely on toxicity information from other sources. Other sources include, but are not limited to the following:

- Toxicological Profiles and Minimal Risk Levels. The Agency for Toxic Substances and Disease Registry (ATSDR)—a federal public health agency of the U.S. Department of Health and Human Services— prepares Toxicological Profiles for hazardous substances in response to statutory requirements under the Comprehensive Environmental Response, Compensation, and Liability Act, commonly known as Superfund.[16] ATSDR's Toxicological Profiles typically evaluate three different exposure durations—acute (14 days or less), intermediate (15-364 days), and chronic (365 days or more). According to ATSDR's website, during the development of toxicological profiles, if the agency determines that reliable and sufficient data exist to identify the specific health effects that result from exposure to a hazardous substance, the agency will derive Minimal Risk Levels. Minimal Risk Levels are an ATSDR estimate of daily human exposure to a hazardous substance at or below which that substance will likely not pose a measurable risk of adverse noncancerous effects—such as neurological, respiratory, and reproductive effects—over a specified time period. Minimal Risk Levels are substance-specific estimates, which according to ATSDR's website are intended to serve as screening levels used by ATSDR health assessors and others to identify contaminants and potential health effects that may be of concern at hazardous waste sites. According to ATSDR's website, for non-carcinogens, ATSDR adopted a practice similar to that of EPA's oral reference dose and inhalation reference concentration. Unlike EPA's IRIS Program, however, ATSDR does not develop quantitative cancer toxicity values.[17]

- Provisional Peer Reviewed Toxicity Values (PPRTV). PPRTVs are toxicity values that EPA's National Center for Environmental Assessment ordinarily prepares on an ongoing basis to support cleanup decisions at Superfund sites. PPRTVs are derived for chronic and subchronic exposure durations in instances where IRIS toxicity assessments are not available, and are sometimes derived for subchronic exposure durations when an IRIS toxicity assessment on chronic exposure exists.[18] Also, while PPRTVs receive internal review by EPA scientists and external peer review by independent scientific experts, they differ from IRIS values in that they do not undergo the same rigorous process of peer review and public participation.
- California Environmental Protection Agency (Cal/EPA) Toxicity Assessments. Cal/EPA prepares toxicity assessments, which are peer-reviewed and provide quantitative values for both cancer and noncancer effects. According to Cal/EPA's website, Cal/EPA's Office of Environmental Health Hazard Assessment is responsible for developing and providing managers in state and local government agencies with toxicological and medical information relevant to managing risks and making decisions involving public health. The office develops procedures and practices for performing health risk

assessments for those involved in environmental health issues, including policymakers, businesspeople, members of community groups, news reporters, and others with an interest in the potential health effects of toxic chemicals. Specifically, the office publishes "A Guide to Health Risk Assessment," which outlines in a generalized form the four risk assessment steps in the National Academies model described above.

EPA's Process for Soliciting Nominations and Selecting Chemicals for New and Updated IRIS Toxicity Assessments

EPA's IRIS Program invites IRIS users to submit nominations for chemicals to be considered for new or updated IRIS toxicity assessments. The IRIS Program solicits nominations from EPA program offices and regions and other federal agencies by issuing a memorandum and solicits nominations from the public by publishing a solicitation in the Federal Register. Generally, the IRIS Program includes a list of criteria that the agency plans to use to prioritize chemicals for selection as part of the solicitation. The IRIS Program included the following criteria in its most recent 2011 nomination solicitation: (1) potential public health impact; (2) EPA statutory, regulatory, or program-specific implementation needs; (3) availability of new scientific information or methodology that might significantly change the current IRIS information; (4) interest to other governmental agencies or the public; (5) availability of other scientific assessment documents that could serve as a basis for development of an IRIS toxicity assessment; and (6) other factors such as widespread exposure to the chemical.[19]

After receiving nominations, IRIS Program staff conducts a preliminary literature search to determine whether there is sufficient information to develop toxicity values for the chemicals nominated. According to IRIS Program officials, the purpose of the literature search is to determine if there is sufficient scientific data from health studies that could be used to develop new IRIS toxicity assessments or update existing assessments. Following the preliminary literature search, the IRIS Program separates the nominated chemicals into two groups: (1) those for which health studies are available and could be used to develop or update an IRIS toxicity assessment and (2) those for which there are not enough data to develop an assessment. Next, according to IRIS Program officials, they provide EPA program offices and regions with an annotated list of chemical nominations that specifies the degree to which health studies are available for each chemical and ask for feedback regarding which chemicals are the highest priorities. After considering such feedback, the IRIS Program selects chemicals for new or updated IRIS toxicity assessments. The IRIS chemical nomination and selection process culminates in the publication in the Federal Register of the IRIS agenda— which contains, among other things, a list of chemicals for which the IRIS Program intends to initiate IRIS toxicity assessments, as well as a list of ongoing IRIS toxicity assessments. For example, the most recent IRIS agenda, published in May 2012, lists 15 IRIS toxicity assessments—with planned start dates ranging from fiscal year 2012 to fiscal year 2014—as well as a list of the 52 IRIS toxicity assessments that were already under way.[20] According to the IRIS agenda, EPA considers its own resources and the availability of guidance, guidelines, and policy decisions in deciding when to start assessments for the selected chemicals.

EPA'S MOST RECENT EVALUATION OF DEMAND FOR IRIS TOXICITY ASSESSMENTS IS A DECADE OLD

EPA has not done a recent evaluation of demand for IRIS toxicity assessments with input from users inside and outside the agency. EPA conducted its most recent evaluation of demand for IRIS toxicity assessments in September 2003, which included input from users inside and outside EPA, but it has not performed a similar review since that time. Without a clear understanding of current demand for IRIS toxicity assessments, EPA cannot adequately measure the program's performance; effectively determine the number of IRIS assessments required to meet the statutory, regulatory, and programmatic needs of IRIS users; or know the extent to which unmet demand exists.

EPA conducted its last evaluation of demand in 2003 at the request of Congress. In September 2000, due to concerns that EPA and state regulators were relying on potentially outdated scientific information, the Senate Committee on Appropriations requested that EPA conduct a needs assessment with public input to determine the need for increasing the annual rate of new and updated IRIS toxicity assessments.[21] In response to the Senate request, EPA conducted a needs assessment— the results of which are discussed in its September 2003 report, *Needs Assessment for U.S. EPA's Integrated Risk Information System*.[22] According to the report, EPA estimated that 50 new or updated IRIS toxicity assessments a year were needed to meet user needs. Specifically, EPA estimated that completing 50 assessments annually would allow the agency to routinely update existing toxicity assessments in the IRIS database, as well as respond to immediate user needs for new or updated assessments.[23] This estimate, according to the report, was based on EPA's past experience soliciting nominations from EPA program offices and regions and information obtained through a July 2001 query of EPA program offices and regions and the public that requested information on which chemicals they considered priorities for assessment.[24] However, based on our review of the report, we did not find sufficient support for the estimate. Specifically, the report did not describe how EPA's past experience or its 2001 query were used to derive the report's estimate that about 50 IRIS toxicity assessments per year were needed to meet demand. In addition, the report stated that because EPA received a small number of responses to the agency's 2001 query, it is not clear if the responses received are necessarily representative of the broad range of IRIS users.

Although EPA's 2003 needs assessment is a decade old, EPA officials told us that the agency does not currently have plans to perform another evaluation of demand for the IRIS Program and that, due to changing conditions over the last 10 years the 2003 evaluation was not applicable to current conditions. IRIS officials stated that the IRIS Program's primary mechanism for monitoring the needs of EPA's program and regional offices at present is to perform outreach with EPA program offices and regions, such as holding quarterly meetings with program office representatives from each EPA program office and holding internal scoping meetings with representatives from EPA program offices and regions for certain chemicals. In response to our questions regarding current demand, IRIS Program officials told us that the annual need for IRIS toxicity assessments may likely be in the hundreds, though officials did not describe how they derived this number.

We have previously reported on the need for EPA to comprehensively analyze its workload and workforce to effectively carry out its strategic goals and objectives. Specifically, in July 2011, we reported that the agency did not have a workload analysis to help determine the optimal numbers and distribution of staff among its laboratory enterprise—which is responsible for providing the scientific research, technical support, and analytical services that underpin its policies and regulations.[25] In addition, we have previously reported that, in developing new initiatives, agencies can benefit from following leading practices for strategic planning.[26] Congress enacted the Government Performance and Results Act (GPRA) in 1993 to improve efficiency and accountability of federal programs.[27] We have reported that these requirements also can serve as leading practices for strategic planning at lower levels within federal agencies, such as planning for individual divisions, programs, or initiatives.[28] Of these leading practices, it is particularly important for agencies to define strategies that address management challenges that threaten their ability to meet long-term goals—including a description of the resources needed to meet established goals. Without an evaluation of current demand for IRIS toxicity assessments that takes into account resource constraints, the IRIS Program risks not being able to develop a plan that lays out realistic goals based on current conditions.

EPA'S CHEMICAL NOMINATION AND SELECTION PROCESS MAY NOT ACCURATELY REFLECT CURRENT DEMAND FOR IRIS TOXICITY ASSESSMENTS

The IRIS Program's chemical nomination and selection process, which the agency uses to gauge interest in the IRIS Program from users inside and outside of EPA, may not accurately reflect current demand for IRIS toxicity assessments. Our analysis of IRIS Program data indicates that the IRIS Program received nominations for 75 chemicals from EPA and non-EPA IRIS users in response to its most recent 2011 nomination period. However, the 75 chemicals received for the 2011 nomination period may not accurately reflect current demand for IRIS toxicity assessments. As about 1,000 new chemicals are listed for commercial use each year, demand for IRIS toxicity assessments is potentially very high, but the number of chemicals nominated may either overstate or understate actual demand. For example, it is not clear how many chemicals IRIS users did not nominate due to concerns that the IRIS toxicity assessment would not be completed in a timely manner. Officials from EPA's Office of Water told us that even though they may need an IRIS toxicity assessment, they sometimes develop their own chemical toxicity assessments to meet their urgent or time-critical needs, such as meeting statutory deadlines.

Also, given the long-standing challenges the IRIS Program has had in routinely starting new assessments, according to some EPA IRIS users, they chose not to nominate new chemicals for assessment and instead nominated chemicals that were already listed on the IRIS agenda as under way. For example, according to officials from EPA's Office of Solid Waste and Emergency Response, due to the large number of chemicals already listed on the IRIS agenda and the IRIS Program's limited resources, in some cases, they reiterated support for chemicals that were already listed on the agenda as under way rather than nominate new chemicals. IRIS Program officials told us that, although EPA program offices and regions and other IRIS users would like to see the IRIS Program produce more IRIS toxicity assessments each year, current resources constrain the speed at which the IRIS Program can complete them. For example, EPA issued 4 IRIS toxicity assessments in fiscal year 2012 (see fig. 2). EPA has issued from 2 to 11 IRIS toxicity assessments annually since fiscal year 2002.

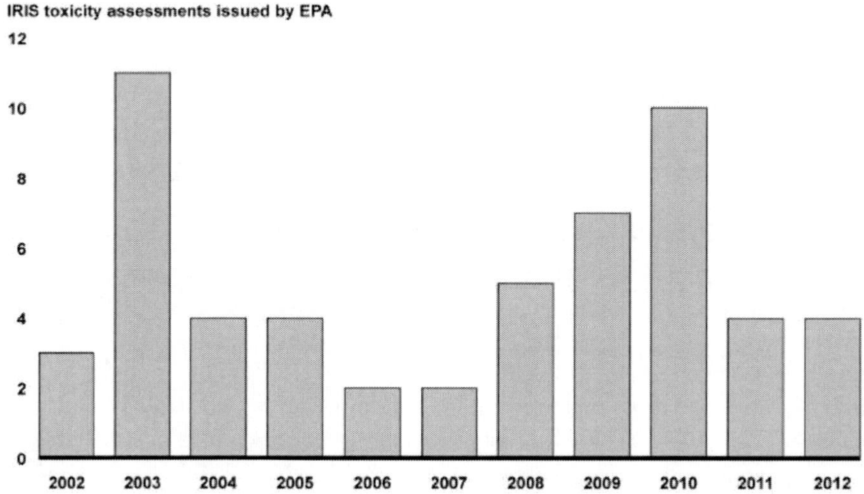

Source: GAO analysis of EPA data.

Note: EPA's Integrated Risk Information System (IRIS) contains toxicity assessments that provide fundamental scientific information EPA needs for assessing the risks chemicals pose to human health (i.e., develop human health risk assessments).

Figure 2. Number of IRIS Toxicity Assessments EPA Issued, Fiscal Years 2002 to 2012.

In addition, as we reported in December 2011, EPA fell short of its productivity goal in fiscal year 2011—completing 4 of the 20 IRIS toxicity assessments it had originally planned to complete that year.[29] Similarly, in completing 4 IRIS toxicity assessments in fiscal year 2012, EPA fell short of its fiscal year 2012 goal of completing 40 assessments. Given the challenges the IRIS Program has experienced issuing IRIS assessments, for many chemicals, the IRIS Program is not fulfilling the goal of providing a common scientific foundation for decision making within EPA programs—which is the stated purpose of an IRIS toxicity assessment according to EPA's strategic planning document for human health risk assessment research.[30] Instead, because the IRIS Program has been unable to keep up with demand for IRIS toxicity assessments, the agency had to prioritize its selection of chemicals for IRIS toxicity assessment.

Furthermore, EPA has not clearly articulated under what circumstances IRIS toxicity assessments are not needed, and the IRIS Program's process for prioritizing chemicals does not provide clarity regarding why specific chemicals are selected for assessment and others are not.[31] According to IRIS Program officials, some chemicals may not need IRIS toxicity assessments. While the IRIS Program has developed criteria that are used to prioritize its selection of chemicals for IRIS toxicity assessment, it is not clear how it applies these criteria—including how it determines the circumstances under which program offices and regions may or may not need an IRIS toxicity assessment. As discussed earlier, the IRIS Program published its chemical selection criteria when it solicited nominations for IRIS toxicity assessment for its 2011 nomination period. However, in announcing that it had selected 15 IRIS toxicity assessments in its 2012 IRIS agenda, the IRIS Program did not explain and has not published information on, how the agency applied its selection criteria.

OMB's implementing guidance for internal control requirements for federal agencies emphasizes the need for agencies to develop policies that ensure the effectiveness and efficiency of their operations and, as part of that, emphasizes that information related to guidance should be communicated to relevant personnel at all levels within an organization and outside the agency in a relevant, reliable, and timely manner.[32]

In August 2012, IRIS Program officials told us that they were working to develop a better description of the nomination and selection process that will clarify how the agency applied the six criteria but, as of March 2013, had not done so. Consequently, for the chemicals that were nominated during the most recent 2011 nomination period, but not selected, it is not clear how many, if any, were excluded from consideration because they did not meet the IRIS Program's selection criteria, because the IRIS Program determined that an IRIS toxicity assessment was not needed— or, alternatively, if they were not selected because of resource constraints or other reasons.

EPA HAS NOT IMPLEMENTED AN AGENCYWIDE STRATEGY FOR ADDRESSING UNMET NEEDS FOR IRIS TOXICITY ASSESSMENTS

EPA has not implemented an agencywide strategy for addressing the unmet needs of EPA program offices and regions when IRIS toxicity assessments are not available, applicable, or current. Specifically, EPA does not have (1) a strategy for identifying and filling data gaps that would enable it to conduct IRIS toxicity assessments for nominated chemicals that were not selected for IRIS toxicity assessment due to insufficient data and (2) agencywide guidance for addressing unmet needs when IRIS toxicity assessments are not available, applicable, or current—which is consistent with findings reported recently by EPA's Inspector General and Science Advisory Board.

EPA Does Not Have a Strategy for Identifying and Filling Data Gaps

EPA does not have a strategy for identifying and filling data gaps that would enable it to conduct IRIS toxicity assessments for nominated chemicals that were not selected for IRIS toxicity assessment because of insufficient scientific data from health studies. As discussed earlier, as part of the IRIS chemical nomination and selection process, IRIS Program officials separate nominated chemicals into two groups: (1) those for which sufficient scientific data from health studies exist that could be used to develop or update an IRIS toxicity assessment and (2) those for which sufficient data do not exist for developing an assessment. For example, as a part of its most recent 2011 nomination period, the IRIS Program dropped 11 of the 75 chemicals nominated from consideration because sufficient scientific data from health studies were not available to develop an IRIS toxicity assessment.[33] One of the chemicals dropped from consideration due to insufficient data was nominated in 2005, 2007, and 2011.[34] The chemical—iso-octane, or 2,2,4-trimethylpentane, which is a constituent of motor fuels—was nominated, according to officials with EPA's Office of Underground Storage Tanks, within the Office of Solid Waste and Emergency Response, so that the office can determine appropriate cleanup levels for leaking underground storage tank sites.[35] Moreover, Section 1505 of the Energy Policy Act of 2005 directed the

EPA Administrator to, among other things, conduct a study on the effects on public health (e.g., the effects on children, pregnant women, minority or low-income communities, and other sensitive populations) of increased use of iso-octane and six other fuel additives as substitutes for methyl tertiary butyl ether (MTBE).[36]

While the IRIS Program prepared and issued an IRIS toxicity assessment that contained the qualitative hazard identification description of isooctane in 2007, it was unable to derive quantitative IRIS values due to insufficient data on the chemical's health effects in humans. According to EPA's 2007 IRIS assessment of iso-octane, the IRIS Program did not develop quantitative estimates of noncancer and cancer risks because the studies needed to support such estimates were not available.[37] Consequently, EPA's Office of Underground Storage Tanks nominated iso-octane again in 2011. In response, according to IRIS Program officials, the IRIS Program evaluated the literature since the 2007 IRIS toxicity assessment was completed and determined that no new studies were available to support development of quantitative IRIS values. Therefore, iso-octane was not considered for an IRIS toxicity assessment in 2011. According to officials with the Office of Underground Storage Tanks, they meet with IRIS Program officials regularly, and the IRIS Program is aware of their need for IRIS toxicity assessments related to these chemicals. However, should officials with the Office of Underground Storage Tanks nominate iso-octane again; EPA cannot ensure that the data needed to prepare an IRIS toxicity assessment that includes quantitative IRIS values will be available and thus, allow EPA to address this unmet need. Without quantitative IRIS toxicity values for these chemicals, it is unclear how EPA will conduct a study of these chemicals on the effects on public health as required by the Energy Policy Act of 2005.

IRIS Program officials told us that no mechanism exists for EPA to ensure that chemicals without sufficient health studies during one nomination period will have those data gaps filled by the next nomination period or even the one after that. As discussed earlier, as part of its solicitation process, the IRIS Program circulates an annotated list of chemical nominations that specifies the degree to which health studies are available for each chemical. However, IRIS Program officials told us that they do not have a process in place for filling research gaps and acknowledged that better coordination across EPA offices and with other federal research agencies, such as the Department of Health and Human Services' National Toxicology Program, could help address this issue. 38 Other agencies within the Department of Health and Human Services may also be a potential research source. Without such research—which is necessary to fill data gaps needed to develop IRIS toxicity assessments—the agency will be unable to ensure that it can respond to unmet EPA program offices' and regions' programmatic and public health needs in the future.

EPA Does Not Have Agencywide Guidance for Addressing Unmet Needs

EPA does not have agencywide guidance for addressing the needs of its program offices and regions when IRIS toxicity assessments are not available, applicable, or current. IRIS Program officials told us that, while there is no agencywide guidance, they work with staff from program offices and regions on a case-by-case basis to find alternatives to IRIS toxicity assessments. For example, IRIS Program officials told us that, in some cases, the Superfund Technical Support Center may be able to partially address the needs of the Office of Solid Waste or Emergency Response or regions by summarizing peer reviewed studies.[39] In other cases, they said that they may work with the Office of Solid Waste and Emergency Response, as well as other program offices and regions, to determine if a PPRTV would meet their needs.[40] In 2008, EPA's Board of Scientific Counselors recommended that EPA consider using PPRTVs as an interim measure to meet its needs for some chemicals, if an IRIS toxicity assessment was not available, and recommended that well-developed PPRTVs be considered as a source of prioritization in the development of full IRIS documents.[41] However, it is unclear how frequently program offices and regions use PPRTVs to support their statutory, regulatory, or programmatic needs—beyond their use in Superfund risk assessments—because EPA does not collect information on or have agencywide guidance on when a PPRTV, or other toxicity assessment, might be an appropriate alternative to an IRIS toxicity assessment. Without such guidance, EPA cannot ensure that it has a consistent approach for addressing the needs of program offices and regions when IRIS toxicity assessments are not available, applicable, or current. Under federal standards of internal control, agencies are to clearly document in writing internal control in management directives, administrative policies, or operating manuals and have it readily available for examination.[42]

In the absence of agencywide guidance, officials from the three EPA program offices and one region we met with said they used a variety of different approaches to meet their needs when an IRIS toxicity assessment was not available or current. That is, EPA offices operate in much the same way as they operated before the IRIS Program was formed to develop consensus opinions within the agency about the health effects from chronic exposure to chemicals. For example, when IRIS toxicity assessments are not available or current, Office of Solid Waste and Emergency Response officials stated that their office refers to a hierarchy of toxicity values it established to perform human health risk assessments for Superfund sites. In 2003, the Office of Solid Waste and Emergency Response updated this hierarchy, which is intended to help risk assessors identify appropriate sources of toxicology information and lists the sources in order of preference as: (1) IRIS toxicity values, (2) PPRTVs, and (3) other EPA and non-EPA sources of toxicity information—with priority given to those sources of information that are the most current, publicly available, and peer reviewed.[43] Similarly, Region 2 officials told us that they generally follow the Office of Solid Waste and Emergency Response's hierarchy of values to guide their decisions regarding other toxicology sources to meet their needs when an IRIS toxicity assessment is not available. For example, Region 2 officials told us that there are a number of chemicals for which their office has used PPRTVs and values from ATSDR and Cal/EPA to address high–profile issues in their region because an IRIS toxicity assessment was not available. However, the officials noted that ATSDR does not develop cancer values.

According to officials with the Office of Water—which is responsible for implementing, among other mandates, the Clean Water Act and Safe Drinking Water Act—they develop toxicity assessments for chemicals to meet statutory deadlines. For example, under the 1996 amendments to the Safe Drinking Water Act, every 5 years, EPA is to determine for at least five unregulated contaminants, including chemicals, whether regulation is warranted, considering those that present the greatest public health concern. Because of the limited number of IRIS toxicity assessments the IRIS Program can select and develop at one time, the Office of Water created the scheme to prioritize and nominate for IRIS toxicity assessment those chemicals that are the most controversial and high-profile, have a high economic impact, and will take more time and staff to complete. The Office of Water can then, according to officials from that office, develop its own assessments for chemicals that have less controversy surrounding them and take less time and staff to complete in order to meet some of its programmatic needs. Officials from the Office of Water told us that the office develops its own assessments for some chemicals because the IRIS Program would not be able to complete most of the needed toxicity assessments in time to meet the office's statutory deadlines.

Similarly, the Office of Pollution Prevention and Toxics, within the Office of Chemical Safety and Pollution Prevention, has developed its own toxicity assessments. The Office of Pollution Prevention and Toxics is responsible for implementing the Toxic Substances Control Act, which provides EPA with the authority to obtain more information on chemicals and to regulate those chemicals that the agency determines pose unreasonable risks to human health or the environment. In February 2012, the office announced plans to develop risk assessments on 83 chemicals. While the office has not nominated any chemicals for IRIS toxicity assessment over the past three nomination periods through the formal nomination process, according to EPA officials with the office, in developing its risk assessments, it plans to incorporate information from IRIS toxicity assessments to the extent such information is available, recent, and relevant. These officials told us that the risk assessments they are conducting in support of the Toxic Substances Control Act are often based on intermittent exposure to workers and consumers who are subject to chemicals contained in products.[44] However, they also told us that, while the IRIS values contained in the database may not always be applicable, often other data available in the IRIS database are applicable, such as toxicity information for shorter term exposure scenarios that have long-lasting/persistent effects (e.g., development toxicity).[45] In these cases, they said that they have used the hazard and dose-response information described in an IRIS toxicity assessment for a particular chemical to develop their own toxicity assessment. IRIS Program officials said that they are working with the Office of Pollution Prevention and Toxics and other EPA offices to find other options for assessing toxicity, such as PPRTVs, when IRIS toxicity assessments are not available, applicable, or current. While IRIS toxicity assessments may not be applicable in all situations, EPA does not have agencywide guidance that outlines the circumstances under which program offices and regions may or may not need IRIS toxicity assessments, or describes appropriate alternative sources to IRIS toxicity assessments.

Our finding concerning the various approaches EPA program offices and regions use to address their need for toxicity assessments is consistent with findings reported recently by EPA's Inspector General and Science Advisory Board. EPA's Office of the Inspector General conducted a survey of 300 respondents from EPA program offices and regions in January 2013. The survey found that 34 percent of the 300 survey respondents indicated that they had experienced a situation in which they or their team researched a substance that was listed in IRIS but used toxicity values from another source instead of those available in IRIS. Of those respondents, 68 percent indicated that one of their top three reasons for doing so was because the alternate system source is more up-to-date with current scientific practice or other information. Additionally, 28 percent of all survey respondents indicated that they had experienced a situation in which they or their team developed their own toxicity values. However, more than a third of respondents indicated that there were no standard operating procedures or other guidance regarding how to choose a source of toxicity values for their office's work.

EPA's Science Advisory Board has also reported on differences across the agency regarding the use of scientific information for decision making.[46] For example, in July 2012, the Science Advisory Board reported that available resources for developing toxicity assessments, the number of scientific staff engaged in the work, and the institutional and legal framework supporting these assessments differ across the agency.[47] The report also noted that some EPA programs and regions do not have the infrastructure required to generate all assessments needed to support their own activities and that scientists in these offices work within statutory constraints, often on an extremely short timetable and with limited budgets. Within those constraints, according to the report, they either assess available scientific information themselves or rely on the Office of Research and Development, other parts of EPA, or other federal or state agencies for the science assessments needed to support decision making.

We have also reported on EPA's fragmented and largely uncoordinated science activities. Specifically, in July 2011, we reported that EPA had not fully addressed the findings and recommendations of five independent evaluations over the past 20 years regarding long-standing planning, coordination, or leadership issues that hamper the quality, effectiveness, and efficiency of EPA's science activities, including its laboratory operations. We recommended, among other things, that EPA establish a top-level science official with the authority and responsibility to coordinate, oversee, and make management decisions regarding major scientific activities throughout the agency, including the work of all program, regional, and Office of Research and Development laboratories.[48] While EPA agreed with our recommendation, it has not fully implemented it. In particular, while EPA expanded the responsibilities of the agency's science advisor to coordinate, oversee, and make recommendations to EPA's Administrator regarding the agency's major scientific activities, as of March 2013, the agency had not given this official the authority to make management decisions regarding scientific activities across EPA as we recommended. In the absence of such authority, there is no agency mechanism for understanding and addressing the unmet needs for IRIS toxicity assessments. As a result, EPA may not be maximizing its limited resources or addressing the statutory, regulatory, and programmatic needs of EPA program offices and regions in a consistent manner.

CONCLUSION

With tens of thousands of chemicals listed with EPA for commercial use in the United States and about 1,000 new chemicals listed for commercial use each year, demand for IRIS toxicity assessments is potentially very high. EPA's IRIS Program develops new toxicity assessments and, as needed, updates information on existing toxicity assessments contained in the IRIS database. EPA has not evaluated demand for IRIS toxicity assessments with input from users inside and outside the agency since 2003, and although IRIS Program officials recognize that the 2003 estimate does not reflect current conditions, the agency does not plan to perform another evaluation of demand. Without a clear understanding of current demand for IRIS toxicity assessments, EPA cannot measure the program's performance; determine the number of IRIS assessments required to meet the statutory, regulatory, and programmatic needs of IRIS users; or know the extent of unmet demand.

The IRIS Program's chemical nomination and selection process, which the agency uses to gauge interest in the IRIS Program from users inside and outside EPA, may not accurately reflect current demand for IRIS toxicity assessments. For example, it is not clear how many chemicals IRIS users did not nominate due to concerns that the IRIS toxicity assessment would not be completed in a timely manner. Furthermore, EPA has not clearly articulated how the IRIS Program applies the criteria it uses to prioritize the selection of chemicals for IRIS toxicity assessment—including how it determines the circumstances under which program offices and regions may or may not need an IRIS toxicity assessment. Consequently, for chemicals that are nominated, but not selected for IRIS toxicity assessment, it is not clear how many, if any, are excluded from consideration because they do not meet the IRIS Program's selection criteria, because the IRIS Program determined that an IRIS toxicity assessment was not needed—or, alternatively, if they are not selected because of resource constraints or other reasons.

EPA has not implemented an agencywide strategy for addressing the unmet needs of EPA program offices and regions when IRIS toxicity assessments are not available, applicable, or current. Specifically, EPA does not have a strategy for identifying and filling data gaps that would enable it to conduct IRIS toxicity assessments for nominated chemicals that were not selected for assessment due to insufficient data. Because EPA does not have a process in place for identifying and filling research gaps, it is unable to ensure it can respond to any unmet EPA program offices' and regions' programmatic and public health needs in the future. Also, EPA does not have guidance that outlines the circumstances under which program offices and regions may or may not need an IRIS toxicity assessment, or that describes appropriate alternative sources to IRIS toxicity assessments. Without guidance, EPA cannot ensure a consistent approach for addressing the needs of program offices and regions when IRIS toxicity assessments are not available, applicable, or current.

RECOMMENDATIONS FOR EXECUTIVE ACTION

We are making three recommendations to the EPA Administrator.

To ensure that EPA can measure the IRIS program's performance and determine the number of IRIS toxicity assessments required to meet the statutory, regulatory, and programmatic needs of IRIS users, we recommend that the EPA Administrator direct the Office of Research and Development to implement the following two actions without impeding the progress of ongoing assessments:

- Identify and evaluate demand for the IRIS Program to determine the number of IRIS toxicity assessments and resources required to meet users' needs.
- Document how EPA applies its IRIS toxicity assessment selection criteria, including the circumstances under which program offices and regions may or may not need an IRIS toxicity assessment.

To ensure that EPA maximizes its limited resources and addresses the statutory, regulatory, and programmatic needs of EPA program offices and regions when IRIS toxicity assessments are not available, we recommend that the EPA Administrator direct the Deputy Administrator, in coordination with EPA's Science Advisor, to implement the following action:

- Once demand for the IRIS Program is determined, develop an agencywide strategy to address the unmet needs of EPA program offices and regions that includes, at a minimum:

- coordination across EPA offices and with other federal research agencies to help identify and fill data gaps that preclude the agency from conducting IRIS toxicity assessments, and
- guidance that describes alternative sources of toxicity information and when it would be appropriate to use them when IRIS values are not available, applicable, or current.

AGENCY COMMENTS AND OUR EVALUATION

We provided a draft of this report to EPA for its review and comment.

In its written comments, EPA agreed with our findings and two of our recommendations and partially agreed with our third recommendation. Specifically, EPA agreed with our recommendations that the Office of Research and Development (1) identify and evaluate demand for the IRIS Program to determine the number of IRIS toxicity assessments and resources required to meet users' needs and (2) document how EPA applies its IRIS toxicity assessment selection criteria, including the circumstances under which program offices and regions may or may not need an IRIS toxicity assessment. In its written comments, EPA stated that the Office of Research and Development this year will evaluate the potential future demand for IRIS toxicity assessments and the resources required to meet that demand. EPA also stated that it will better describe for internal and external stakeholders and the public the nomination and selection process for chemicals for IRIS toxicity assessments, including the rationale for not selecting nominated chemicals for IRIS assessment.

With respect to our third recommendation, that EPA develop an agencywide strategy to address unmet need for IRIS toxicity assessments, in its written comments, EPA requested that we provide additional clarification and consider refining our recommendation. Specifically, EPA stated that it understood and supported the goal of developing an agencywide strategy to help identify and fill data gaps that preclude the agency from conducting IRIS toxicity assessments, but urged us to clarify more precisely the extent to which it must rely on others to conduct research to fill data gaps on IRIS chemicals. As we note in the report, IRIS Program officials told us that they do not have a process in place for filling research gaps and acknowledged that better coordination across EPA offices and with other federal research agencies, such as the Department of Health and Human Services' National Toxicology Program, could help address this issue. We acknowledge that EPA has limited resources, which may preclude the agency from making substantial investments in research into how individual chemicals affect human health. As such, to ensure that the agency maximizes its limited resources, we have recommended that EPA develop a strategy to coordinate with other federal research agencies to help identify and fill data gaps. In this context, EPA acknowledged that it must look to other federal agencies, academic institutions, and chemical product producers to fund research into how chemicals affect human health, as we have recommended. EPA also stated, in its written comments, that the agency can and will do a more effective job to make data needs known to relevant federal agencies and nonfederal organizations that either fund or conduct chemical research.

Also regarding our third recommendation that EPA develop an agencywide strategy to address unmet needs for IRIS toxicity assessments, based on technical comments provided by EPA officials prior to receiving the agency's letter dated April 16, 2013, we refined the wording of our third recommendation. The original text recommended that EPA develop guidance that describes alternative sources of toxicity information and *procedures for preparing toxicity assessments* when IRIS values are not available, applicable, or current. We refined the wording of this recommendation to read: guidance that describes alternative sources of toxicity information and *when it would be appropriate to use them* when IRIS values are not available, applicable, or current. The revised language more accurately reflects the intent of our recommendation. In addition, in its written comments, EPA stated that it understood our interest in the agency developing guidance that describes alternative sources of toxicity information and agreed that such guidance might be helpful. However, EPA stated that the development of such guidance is best left to individual EPA programs. We disagree. In the absence of agencywide guidance that addresses unmet demand for IRIS assessments, EPA offices operate in much the same way they operated before the IRIS Program was formed to develop consensus opinions within the agency about the health effects from chronic exposure to chemicals. As we note in this report, we have previously reported on EPA's fragmented and largely uncoordinated science activities and recommended, among other things, that EPA establish a top-level science official with the authority and responsibility to coordinate, oversee, and make management decisions regarding major scientific activities throughout the agency.[49] Consistent with our prior report and recommendation, we believe that guidance regarding major scientific activities should also come from a top-level science official. However, as we note in our current report, EPA has not provided its Science Advisor with the authority to make management decisions regarding scientific activities across EPA as we previously recommended. Therefore, we believe that agencywide guidance should come from EPA's Deputy Administrator in coordination with EPA's Science Advisor.

Sincerely yours,

J. Alfredo Gómez
Director, Natural Resources and Environment

APPENDIX I: SCOPE AND METHODOLOGY

To determine the extent to which the Environmental Protection Agency (EPA) has evaluated demand for IRIS toxicity assessments, we reviewed EPA's 2003 evaluation of demand for IRIS toxicity assessments. We also interviewed IRIS Program officials to determine whether they had conducted other evaluations of demand since 2003, how they derived the 2003 estimate, and whether that estimate reflects current conditions. Because the 2003 evaluation did not provide sufficient information on its methodology, we were unable to fully assess its estimate. We corroborated EPA officials' assertion that the 2003 assessment no longer reflects current conditions based on our understanding of the IRIS Program. As we discussed earlier, the importance of EPA's IRIS Program has increased over time as EPA program offices and regions have increasingly relied on IRIS toxicity assessments in making environmental protection and risk management decisions. In addition, as about 1,000 new chemicals are listed for commercial use each year, potential for changes in demand over time are likely.

To determine the extent to which EPA's process for nominating and selecting chemical for IRIS toxicity assessment accurately reflects current demand, we reviewed data provided by the IRIS Program and from the IRIS Program's website on the number of IRIS toxicity assessments it completed annually from fiscal years 2002 through 2012. In addition, we analyzed all chemical nomination forms submitted by EPA program offices and regions to the IRIS Program from 2005, 2007 and 2011— which were the last three times that EPA solicited nominations for new and updated IRIS toxicity assessments.[1] For additional perspective on user needs, we reviewed non-EPA IRIS users' chemical nomination forms from 2011. To select and count the number of nominations, two analysts reviewed information EPA provided us to determine which documents to include in our analysis. We used the following inclusion and exclusion criteria to determine which documents to include in our analysis:

- Documents labeled as being a nomination were included while documents labeled as being another document were excluded. For example, some documents were the IRIS Program's request for nominations or Federal Register Notices, which were not included in our analysis. In other instances, a nominating entity indicated that the chemicals were being prioritized and not nominated (i.e., they clearly stated "this is a list of our priorities"), and were not included in our analysis.
- We included individual nomination sheets for our analysis, but we did not include nomination cover sheets or separate documents that IRIS Program officials sent us separately from the individual nomination sheets.
- We did not include the nomination form if the nominating entity indicated on the form that there was no new nomination and instead was a reiteration of support for a previous nomination.

In some instances, EPA program offices and regions nominated two chemicals on one nomination form or listed two chemicals together. Chemicals were counted as a single nomination if they had the same Chemical Abstracts Service (CAS) registry number and as separate nominations if they had different CAS registry numbers. According to the CAS registry website, the CAS registry is the most authoritative collection of disclosed chemical substance information, containing more than 70 million organic and inorganic substances and 64 million sequences. In cases where it was not clear to both analysts whether to include a document in our analysis, IRIS Program officials provided confirmation. While we used this methodology to determine the number of nominations in the 2005, 2007, and 2011 nomination periods, using a different methodology might result in a different number of nominations. As about 1,000 new chemicals are listed for commercial use each year, the chemicals nominated may either overstate or understate actual demand. In addition, we reviewed the IRIS Program's processes for soliciting nominations and selecting chemicals for IRIS toxicity assessments, including the agency's selection criteria. We also reviewed agency guidance and interviewed IRIS Program officials to better understand the chemical nomination and selection process.

To determine the extent to which EPA has implemented a strategy for addressing any unmet needs of EPA program offices and regions when IRIS toxicity assessments are not available, applicable, or current, we reviewed the IRIS Program's efforts to analyze IRIS user chemical nominations. For context, we interviewed officials from EPA's National Center for Environmental Assessment, which manages the IRIS Program and develops IRIS toxicity assessments. For additional perspective, we interviewed officials using a standard set of questions from a nonprobability sample of three EPA program offices and one region: the Office of the Administrator, the Office of Water, the Office of Solid Waste and Emergency Response, and EPA's Region 2.[2] We selected these program offices and region because they submitted 78 percent of the chemical nominations to the IRIS Program during the period we reviewed—2005, 2007, and 2011. These offices and region ranked the highest in terms of the number of chemical nominations submitted, and, in some cases, nominated chemicals more than once during different nomination years. Because this is a nonprobability sample, it is not generalizable to all EPA program offices and regions, but it can provide illustrative examples of the experience of those EPA program offices and one region that nominated 78 percent of chemicals for IRIS toxicity assessment during the period we reviewed. For example, we received information from officials from these offices about how EPA program offices and one region nominate chemicals for IRIS toxicity assessment, how the IRIS Program meets the needs of these offices and region over the course of nomination periods, and what alternative toxicity assessments these offices and region turn to when IRIS toxicity assessments are not available. For a summary of approaches used by selected EPA program offices and regions to address their IRIS toxicity assessment needs, see appendix II.

Separately from our nonprobability sample, we also interviewed officials from the Office of Pollution Prevention and Toxics, within the Office of Chemical Safety and Pollution Prevention, because it did not nominate any chemicals for IRIS toxicity assessment for any of the last three nomination periods. We did not evaluate the scientific content or quality of IRIS toxicity assessments.

We conducted this performance audit from April 2012 to May 2013 in accordance with generally accepted government auditing standards. Those standards require that we plan and perform the audit to obtain sufficient, appropriate evidence to provide a reasonable basis for our findings and conclusions based on our audit objectives. We believe that the evidence obtained provides a reasonable basis for our findings and conclusions based on our audit objectives.

APPENDIX II: SUMMARY OF APPROACHES USED BY SELECT EPA PROGRAM OFFICES AND REGIONS TO ADDRESS TOXICITY ASSESSMENT NEEDS

Office of Solid Waste and Emergency Response

According to officials at the Office of Solid Waste and Emergency Response, the office nominated 18 chemicals over the course of the 2005, 2007, and 2011 nomination periods.[1] The Office of Solid Waste and Emergency Response provides policy, guidance, and direction for the agency's emergency response and waste programs, including managing the Superfund Program, which responds to abandoned and active hazardous waste sites and accidental oil and chemical releases. IRIS toxicity assessments are used by the Office of Solid Waste and Emergency Response to, among other things, support mandated regulatory actions. For example, the Office of Underground Storage Tanks, within the Office of Solid Waste and Emergency Response, submitted chemicals during the 2011 nomination period to support the requirement under Section 1505 of the Energy Policy Act of 2005 that the EPA Administrator conduct a study on the effects on public health of increased use of iso-octane and six other fuel additives as substitutes for methyl tertiary butyl ether (MTBE).[2] When IRIS toxicity assessments are not available or current, Office of Solid Waste and Emergency Response officials stated they rely on other toxicity values to meet their programmatic needs. For example, officials at the Office of Underground Storage Tanks stated that, in the absence of IRIS values, states must resort to other sources for toxicological information, and this can lead to inconsistencies state-to-state. Officials also stated that, when an IRIS toxicity assessment is not available, the office refers to a hierarchy of toxicity values to be used in performing human health risk assessments for Superfund sites. In 2003, the Office of Solid Waste and Emergency Response updated this hierarchy, which is intended to help risk assessors identify appropriate sources of toxicology information and lists the sources as: (1) IRIS toxicity values, (2) Provisional Peer Reviewed Toxicity Values (PPRTVs), and (3) other EPA and non-EPA sources of toxicity information—with priority given to those sources of information that are the most current, publicly available, and peer reviewed. Such values include the Agency for Toxic Substances and Disease Registry (ATSDR) Minimal Risk Levels and California Environmental Protection Agency (Cal/EPA) toxicity values. Although developed specifically for the Superfund Program, officials stated that this guidance is generally used by all suboffices within the Office of Solid Waste and Emergency Response.

EPA Region 2

According to Region 2 officials, the region nominated 22 chemicals over the course of the 2005, 2007, and 2011 nomination periods. EPA's Region 2 serves New Jersey, New York, Puerto Rico, the U.S. Virgin Islands, and eight tribal nations. Region 2 nominates chemicals for IRIS assessment on behalf of risk assessors throughout their region—that is, EPA and officials throughout the region, primarily at Superfund sites, that evaluate chemical risks. For example, Region 2 stated in its chemical nomination form for the 2011 nomination period that it needed IRIS toxicity assessments to support cleanup decisions for chemicals present in residential properties and in groundwater. Region 2 officials indicated that, when IRIS toxicity values are not available, they may rely on other toxicity values to meet their programmatic needs and follow the Office of Solid Waste and Emergency Response's hierarchy of values in consultation with the IRIS Program as appropriate.[3] Officials also told us that there are number of high-profile chemicals where the office has used PPRTVs and values from ATSDR and Cal/EPA, because an IRIS toxicity assessment is not available. However, officials noted that ATSDR does not develop cancer values.

Office of Water

According to Office of Water officials, the office nominated 23 chemicals over the course of the 2005, 2007, and 2011 nomination periods. EPA's Office of Water is responsible for drinking water safety, and it restores and maintains oceans, watersheds, and their aquatic ecosystems. The Office of Water is responsible for implementing, among other mandates, the Clean Water Act and Safe Drinking Water Act. For example, in its chemical nomination form for the 2011 nomination period, the Office of Water stated that it needed IRIS toxicity assessments to develop regulations. The Office of Water develops assessments for chemicals it needs to meet statutory deadlines. Because of the limited number of IRIS toxicity assessments the IRIS Program can select and develop at one time, the Office of Water created the scheme to prioritize those chemicals that are the most controversial and high-profile, have a high economic impact, and will take more staff and time to complete. The Office of Water can then develop its own assessments for chemicals that have less controversy surrounding them and take less time and staff to complete in order to meet some of its programmatic needs. Officials stated that developing the office's own assessments for some chemicals was based on the reality that the IRIS Program would not complete most of the needed toxicity assessments in time to meet the office's statutory deadlines.

Office of the Administrator

According to officials at the Office of the Administrator, the office nominated 26 chemicals over the course of the 2005, 2007, and 2011 nomination periods.[4] EPA's Office of the Administrator provides executive and logistical support for the EPA Administrator. The office supports the leadership of EPA's programs and activities to protect human health and the environment. An official from the Office of Policy, within the Office of the Administrator, stated that rationales for nominating chemicals varied widely—for example, the increasing or widespread exposure to a chemical or the availability of new data to develop a new or update an existing IRIS toxicity assessment. The official noted that the Office of Policy's programmatic needs differed from other EPA offices' needs in that it does not develop regulations or risk assessments. Instead, the office provides assistance for other EPA offices' assessments and reviews assessments that other offices perform. In the absence of an IRIS toxicity assessment, the Office of Policy relied on the original literature, review articles, and assessments prepared by other agencies. Such values include the ATSDR Minimal Risk Levels and Cal/EPA toxicity values. An official with the Office of Children's Health Protection, also within the Office of the Administrator, stated that most of its nominations were for chemicals that were under the Toxic Substances Control Act or the Safe Drinking Water Act and were based on children's health concerns. The official stated that, in the absence of an IRIS toxicity assessment, the Office of Children's Health Protection goes directly to the literature or work done by other government agencies or programs, such as the National Toxicology Program.

Office of Pollution Prevention and Toxics

The Office of Pollution Prevention and Toxics is responsible for implementing the Toxic Substances Control Act, which provides EPA with the authority to obtain more information on chemicals, and to regulate those chemicals that the agency determines pose unreasonable risks to human health or the environment, announced in February 2012 its plans to develop risk assessments on 83 chemicals. While the office has not nominated any chemicals for IRIS toxicity assessment over the past three nomination periods through the formal nomination process, according to EPA officials with the Office of Pollution Prevention and Toxics, in developing its risk assessments, it plans to incorporate information from IRIS toxicity assessments to the extent such information is available, recent, and relevant. Officials at the Office of Pollution Prevention and Toxics, and senior staff of the Office of Research and Development, which houses the IRIS Program, have compared the list of existing and ongoing IRIS toxicity assessments in order to share relevant literature and hazard reviews for upcoming risk assessments related to the Toxic Substances Control Act. These officials told us that the risk assessments they are conducting in support of the Toxic Substances Control Act are often based on intermittent exposure to workers and consumers who are subject to chemicals contained in products. However, they also told us that, while the IRIS values contained in the database may not always be applicable, often other data available in the IRIS database are applicable, such as toxicity information for shorter term exposure scenarios that have long-lasting/persistent effects (e.g., development toxicity). In these cases, they said that they have used the hazard and dose response information described in an IRIS toxicity assessment for a particular chemical to develop their own toxicity assessment.

In addition, according to IRIS program officials, they have compared the list of chemicals for which the Office of Pollution Prevention and Toxics plans to conduct risk assessments with the list of existing and ongoing toxicity assessments and shared relevant literature and hazard reviews. IRIS program officials also said that the Office of Pollution Prevention and Toxics participates with other EPA offices in prioritizing their needs for ongoing IRIS toxicity assessments. IRIS toxicity values are generally used to estimate risks associated with continuous exposures to a pollutant in the air or water. In most cases, according to IRIS Program officials, the information used to develop the dose-response assessments is based on intermittent exposures to workers or animals in a controlled environment, and IRIS assessments include an adjustment to continuous exposure in the derivation of toxicity values. IRIS Program officials said that they are working with the Office of Pollution Prevention and Toxics and other EPA offices to find other options for assessing toxicity, such as PPRTVs, when IRIS toxicity assessments are not available, applicable, or current.

End Notes

[1] Under its IRIS Program, EPA (1) identifies a chemical's toxicity, or hazardous properties, which are the potential noncancer and cancer human health effects of exposure to a chemical, and (2) assesses the dose-response relationship between exposure to a chemical and the resultant health effects, which describes the magnitude of hazard for potential noncancer effects and increased cancer risk.

[2] A human health risk assessment characterizes the nature and magnitude of health risks to humans from exposure to chemical contaminants that may be present in the environment. IRIS toxicity assessments are used along with other information to prepare human health risk assessments. Toxicity represents the degree to which a chemical is harmful. In this report, the terms toxicity and hazard are used synonymously.

[3] According to IRIS Program officials, new assessments are added to the IRIS database after an evaluation of the available scientific literature, and existing assessments are revised based on an evaluation of new studies that have been published since the original assessment was completed.

[4] GAO, *Chemical Assessments: Low Productivity and New Interagency Review Process Limit the Usefulness and Credibility of EPA's Integrated Risk Information System,* GAO-08-440 (Washington, D.C.: Mar. 7, 2008).

[5] GAO, *High-Risk Series: An Update*, GAO-09-271 (Washington, D.C.: Jan. 22, 2009). This high-risk area addresses EPA's implementation of the IRIS Program, as well as implementation of the Toxic Substances Control Act.

[6] GAO, *Chemical Assessments: Challenges Remain with EPA's Integrated Risk Information System Program*, GAO-12-42 (Washington, D.C.: Dec. 9, 2011).

[7] For purposes of this report, "IRIS users" represent EPA program offices and regions and non-EPA entities that submit chemical nominations to the IRIS Program. Such non-EPA entities include other federal agencies, White House offices, and the public.

[8] According to IRIS Program documents, the IRIS Program solicits opinions from EPA offices, other federal agencies, and the public on an annual basis. However, since 2004, the IRIS Program has solicited nominations from IRIS users three times: in 2004, 2006, and 2010. In their nomination forms, IRIS users send EPA the names of chemicals and the reasons for requesting that IRIS toxicity assessments be developed or updated, among other information. In 2004, the IRIS Program solicited nominations only from EPA program offices and regions. In 2006 and 2010, the IRIS Program solicited nominations from all IRIS users. The IRIS Program refers to these nomination periods as being for fiscal years 2005, 2007, and 2011, which are the years we also refer to in this report.

[9] We interviewed officials both at the Office of Policy and the Office of Children's Health Protection, which are suboffices within the Office of the Administrator. We also received a

written response from the Office of Underground Storage Tanks, which is a suboffice within the Office of Solid Waste and Emergency Response.

[10] The National Academies comprise four organizations: the National Academy of Sciences, the National Academy of Engineering, the Institute of Medicine, and the National Research Council. EPA uses the model that was initially outlined in the 1983 National Academies' National Research Council, *Risk Assessment in the Federal Government* (commonly known as the Red Book). The National Research Council's subsequent 2009 report presented by the National Academies in *Science and Decisions: Advancing Risk Assessment* (Washington, D.C.: The National Academies Press, 2009) reiterated this process. This publication is also known as the Silver Book.

[11] Exposure represents the magnitude, frequency, and duration of contact with a chemical.

[12] According to IRIS Program officials, one of the following five descriptors is selected based on the extent of human and animal data available: (1) "carcinogenic to humans," (2) "likely to be carcinogenic to humans," (3) "suggestive evidence of carcinogenic potential," (4) "inadequate information to assess carcinogenic potential," or (5) "not likely to be carcinogenic to humans."

[13] OMB, Fiscal Year 2006 Program Assessment Rating Tool (PART) assessment of EPA's Human Health Risk Assessment Program.

[14] EPA Office of Research and Development, *Human Health Risk Assessment Strategic Research Action Plan 2012-2016*, EPA 601/R-12/007, June 2012.

[15] EPA decides the type of independent peer review an IRIS toxicity assessment will undergo. The peer reviews are conducted by (1) a peer review panel assembled by an EPA contractor, (2) EPA's Science Advisory Board, or (3) the National Academies.

[16] 42 U.S.C. § 9604 (i)(3).

[17] Toxicity values are a numerical expression of a substance's exposure-response relationship that is used in risk assessments.

[18] The specific definition for each exposure duration category may vary depending on the source of the toxicity value being used. In general, subchronic is used to describe periods of repeated exposure by the oral, dermal, or inhalation route for more than 30 days up to approximately 90 days in laboratory animals or 10 percent of the life span in humans.

[19] The IRIS Program also included these criteria, with slight variation, for the 2005 and 2007 nomination solicitations.

[20] Of the 15 IRIS toxicity assessments listed in the 2012 IRIS agenda, 7 were updates of assessments already in the IRIS database. Of the 52 IRIS toxicity assessments that were already under way, 34 were updates of assessments already in the IRIS database.

[21] S. Rep. No. 106-410 at 90 (2000).

[22] EPA, *Needs Assessment for U.S. EPA's Integrated Risk Information System* (Washington, D.C.: September 2003).

[23] For example, the report called for developing a schedule for updating potentially outdated assessments at least once every 10 years and updating some every 5 years.

[24] EPA received responses from 10 EPA program offices, 7 EPA regions, and 22 public and private organizations. Non-EPA respondents included the U.S. Army, state agencies, industries, trade organizations, public interest and nonprofit organizations.

[25] GAO, *Environmental Protection Agency: To Better Fulfill Its Mission, EPA Needs a More Coordinated Approach to Managing Its Laboratories*, GAO-11-347 (Washington, D.C.: July 25, 2011).

[26] GAO, *Environmental Protection: EPA Should Develop a Strategic Plan for Its New Compliance Initiative*, GAO-13-115 (Washington, D.C.: Dec. 10, 2012); GAO,

Environmental Justice: EPA Needs to Take Additional Actions to Help Ensure Effective Implementation, GAO-12-77 (Washington, D.C.: Oct. 6, 2011).

[27] We have reported in the past that, taken together, the strategic planning elements established under GPRA and associated OMB guidance, and practices identified by GAO provide a framework of leading practices in federal strategic planning. For example, see GAO, *Executive Guide: Effectively Implementing the Government Performance and Results Act*, GAO/GGD-96-118 (Washington, D.C.: June 1, 1996); GAO, *Tax Administration: IRS Needs to Further Refine Its Tax Filing Season Performance Measures*, GAO-03-143 (Washington, D.C.: Nov. 22, 2002); and GAO, *Managing for Results: Strengthening Regulatory Agencies' Performance Management Practices*, GAO/GGD-00-10 (Washington, D.C.: Oct. 28, 1999).

[28] Leading practices in federal strategic planning include defining mission and goals, defining strategies that address management challenges and identify resources needed to achieve goals, ensuring leadership involvement and accountability, involving stakeholders, coordinating with other federal agencies, and developing and using performance measures.

[29] GAO-12-42. EPA originally planned to issue 20 IRIS toxicity assessments in fiscal year 2011: 4 were issued, 1 was dropped, 2 were given "TBD"—to be determined—completion dates, and 13 were added to the fiscal year 2012 completion goal, bringing it to 40 assessments.

[30] EPA Office of Research and Development, *Human Health Risk Assessment Strategic Research Action Plan 2012-2016*, EPA 601/R-12/007, June 2012.

[31] As discussed earlier, EPA's IRIS Program was originally created to develop consensus across the Agency regarding health effects resulting from chronic exposures to chemicals. IRIS Program officials have recently told us that IRIS no longer formally requires Agency consensus for its assessments, but consensus is still sought and often achieved through the existing Agency review process.

[32] OMB, *Memorandum to the Chief Financial Officers, Chief Operation Officers, Chief Information Officers, and Program Managers*, December 21, 2004.

[33] As discussed earlier, the IRIS Program has not published information on how the agency applied its selection criteria but, upon request, provided us with information regarding chemicals dropped from consideration due to insufficient data.

[34] 2,2,4-Trimethylpentane (iso-octane).

[35] According to EPA's website, 2,2,4-trimethylpentane is released to the environment through the manufacture, use, and disposal of products associated with the petroleum and gasoline industry. During an accident, 2,2,4-trimethylpentane penetrated the skin of a human which caused necrosis of the skin and tissue in the hand and required surgery. No other information is available on the acute (short-term) effects in humans. Irritation of the lungs, edema, and hemorrhage has been reported in rodents acutely exposed by inhalation and injection. No information is available on the chronic (long-term), reproductive, developmental, or carcinogenic effects of 2,2,4-trimethylpentane in humans. Kidney and liver effects have been observed in rats chronically exposed via gavage (experimentally placing the chemical in the stomach) and inhalation.

[36] 42 U.S.C. § 7545(b)(4). These substances are (1) ethyl tertiary butyl ether (ETBE); (2) tertiary amyl methyl ether (TAME); (3) diisopropyl ether (DIPE); (4) tertiary butyl alcohol (TBA); (5) ethanol; (6) iso-octane (2,2,4-trimethylpentane); and (7) alkylates. According to EPA's website, MTBE is a fuel additive made by combining methanol and isobutylene and has been used since 1979 in the United States as an octane-enhancing replacement for lead. According to the website, the use of MTBE in gasoline sold in the United States has virtually ceased in recent years. According to IRIS Program officials, ETBE, TAME, DIPE,

TBA, and ethanol are currently on the IRIS agenda in the draft development step. The ETBE assessment is being revised because important new data became available after the external peer review. New schedules for ETBE and TBA were posted on the IRIS website in February 2013. Taking into account the complexity of the ethanol dataset, IRIS Program officials told us that they are considering various approaches to conducting the ethanol assessment and will revisit the priority of the ethanol assessment in fiscal year 2013. Alkylates were added to the IRIS agenda in 2007 but were withdrawn in 2012 because there are multiple chemicals in this class, many with limited databases. If individual alkylates with sufficient data to support an IRIS assessment are nominated in the future, the IRIS Program will consider these nominations individually.

[37] According to EPA's 2007 IRIS assessment on iso-octane, no subchronic or chronic oral or inhalation studies were identified that demonstrated a dose-response effect that could be used to determine the noncarcinogenic risk for iso-octane and no studies were available on the carcinogenic effects of iso-octane on which to base a cancer assessment.

[38] The National Toxicology Program was created in 1978 as a cooperative effort to coordinate toxicology testing programs within the federal government, strengthen the science base in toxicology, develop and validate improved testing methods, and to provide information about potentially toxic chemicals to health, regulatory, and research agencies, scientific and medical communities, and the public.

[39] According to officials at the Office of Solid Waste and Emergency Response, summaries of toxicity literature are more useful than no information about toxicity of a particular chemical, but typically the summaries do not completely address the needs of the program offices of regions.

[40] PPRTVs are developed for EPA's Superfund program and, therefore, the Superfund program determines the chemicals that are nominated for development. Other EPA program offices are not included in the nomination process in the same fashion as they are for the IRIS nomination process. However, once a PPRTV is posted to the Office of Solid Waste and Emergency Response's Superfund website, they are publicly available.

[41] EPA Board of Scientific Counselors, *Review of the Office of Research and Development's Human Health Risk Assessment Program at the U.S. Environmental Protection Agency: Final Report*, April 1, 2008.

[42] GAO, *Standards for Internal Control in the Federal Government*, GAO/AIMD-00-21.3.1 (Washington, D.C.: November 1999).

[43] Other non-EPA sources of toxicity information include ATSDR Minimal Risk Levels and Cal/EPA toxicity values. Although the Office of Solid Waste and Emergency Response developed the hierarchy of toxicity values specifically for the Superfund Program, officials stated that the hierarchy is generally used by all suboffices within the Office of Solid Waste and Emergency Response when IRIS toxicity assessments are not available or current.

[44] This focus is consistent with the fact that media-specific environmental laws such as the Clean Air Act and Clean Water Act are available to limit the concentration of contaminants in water or the ambient air, and the Toxic Substances Control Act requires EPA to defer action to such other laws if the agency determines that it can use them to adequately address a given risk. However, information on such continuous exposures is still critical for regulation under the Toxic Substances Control Act. For example, to promulgate a rule under section 6 of the Toxic Substances Control Act, EPA must have information on, among other things, the effects of the chemical on human health and the magnitude of human exposure to the chemical.

[45] IRIS toxicity values are generally used to estimate risks associated with continuous exposures to a pollutant in the air or water. In most cases, the information used to develop the dose-response assessments is based on intermittent exposures to workers or animals in a controlled environment. IRIS assessments include an adjustment to continuous exposure in the derivation of toxicity values.

[46] EPA's Science Advisory Board is a federal advisory committee established by Congress in 1978 with a broad mandate to advise the agency on technical matters.

[47] Science Advisory Board, Science Integration for Decision Making at the U.S. Environmental Protection Agency, EPA-SAB-12-008 (Washington, D.C.: July 6, 2012).

[48] GAO-11-347.

[49] GAO-11-347.

End Notes for Appendix I

[1] According to IRIS Program documents, the IRIS Program solicits opinions from EPA offices, other federal agencies, and the public on an annual basis. However, since 2004, the IRIS Program has solicited nominations from IRIS users three times: in 2004, 2006, and 2010. In 2004, the IRIS Program solicited nominations only from EPA program offices and regions. In 2006 and 2010, the IRIS Program solicited nominations from all IRIS users. The IRIS Program refers to these nomination periods as being for fiscal years 2005, 2007, and 2011, which are the years we also refer to in this report.

[2] We interviewed officials both at the Office of Policy and the Office of Children's Health Protection, which are suboffices within the Office of the Administrator. We also received a written response from the Office of Underground Storage Tanks, which is a suboffice within the Office of Solid Waste and Emergency Response.

End Notes for Appendix II

[1] The Office of Solid Waste and Emergency Response has suboffices including the Office of Underground Storage Tanks, which submitted separate nominations (inclusive of the 18 chemicals) to the IRIS Program.

[2] 42 U.S.C. § 7545(b)(4). These substances are (1) ethyl tertiary butyl ether (ETBE); (2) tertiary amyl methyl ether (TAME); (3) diisopropyl ether (DIPE); (4) tertiary butyl alcohol (TBA); (5) ethanol; (6) iso-octane (2,2,4-trimethylpentane); and (7) alkylates. According to EPA's website, MTBE is a fuel additive made by combining methanol and isobutylene and has been used since 1979 in the United States as an octane-enhancing replacement for lead. According to the website, the use of MTBE in gasoline sold in the United States has virtually ceased in recent years. According to IRIS Program officials, ETBE, TAME, DIPE, TBA, and ethanol are currently on the IRIS agenda in the draft development step. The ETBE assessment is being revised because important new data became available after the external peer review. New schedules for ETBE and TBA were posted on the IRIS website in February 2013. Taking into account the complexity of the ethanol dataset, IRIS Program officials told us that they are considering various approaches to conducting the ethanol assessment and will revisit the priority of the ethanol assessment in fiscal year 2013.

Alkylates were added to the IRIS agenda in 2007 but were withdrawn in 2012 because there are multiple chemicals in this class, many with limited databases. If individual alkylates with sufficient data to support an IRIS assessment are nominated in the future, the IRIS Program will consider these nominations individually.

[3] Region 2 officials stated that in some cases, other organizations such as ATSDR or Cal/EPA may develop a quantitative value before the IRIS toxicity assessment is revised. In that case, the Region would consider the use of the quantitative value based on discussions with the IRIS program.

[4] The Office of the Administrator has suboffices including the Office of Policy and the Office of Children's Health Protection, which each submitted separate nominations (inclusive of the 26 chemicals) to the IRIS Program.

In: Health Assessments …
Editor: Marcus Rowe

ISBN: 978-1-63321-166-7

Chapter 2

EPA'S INTEGRATED RISK INFORMATION SYSTEM PROGRAM PROGRESS REPORT AND REPORT TO CONGRESS*

U.S. Environmental Protection Agency: Office of Research and Development

1. EXECUTIVE SUMMARY

The U.S. Environmental Protection Agency's (EPA) Integrated Risk Information System (IRIS) Program provides health effects information on chemicals to which the public may be exposed, providing a critical part of the scientific foundation for EPA's decisions to protect public health. EPA has made several changes to this important Program over the past few years, streamlining the assessment development process, improving transparency, and creating efficiencies within the Program. In April 2011, the National Research Council (NRC) made several recommendations to EPA for improving the development of IRIS assessments. The NRC's recommendations were focused on the development of draft assessments, and the NRC was clear that their intent was not to delay assessments.

EPA has made progress in implementing these recommendations. Consistent with the advice of the NRC, EPA is implementing these

* This is an edited, reformatted and augmented version of a report released June 2012.

recommendations using a phased approach and is making the most extensive changes to documents that are in the earlier steps of the assessment development process. For assessments that are in the later stages of development, including assessments that have been posted on the IRIS database since the release of the NRC report, EPA is implementing the recommendations as feasible without taking the assessments backwards to earlier steps of the process. Phase 1 of implementing the NRC recommendations has focused on editing and streamlining documents and using more tables, figures, and appendices. EPA is now in Phase 2 of implementing the NRC recommendations and will soon publicly release two draft IRIS assessments that represent a major advancement in implementing the NRC recommendations. EPA is using a new document structure for these draft assessments, including an Executive Summary presenting major conclusions, a Preamble describing methods used to develop the assessment, distinct sections on Hazard Identification and Dose-Response Analysis, and more tables and figures to clearly present data. Additionally as part of Phase **2,** Phase 2, EPA is addressing all of the short-term recommendations provided by the NRC. As part of this effort, EPA will make several changes to IRIS assessments. Highlights include: evaluating and describing the strengths and weaknesses of critical studies in a more uniform way; including toxicity values for multiple effects associated with the chemical, if applicable and where the data allow; routinely considering the use of multiple data sets of combined multiple responses in deriving toxicity values, where appropriate; and evaluating existing guidelines to establish clearer criteria for study selection. Phase 3 of implementation will incorporate the longer-term scientific recommendations made by the NRC.

The U.S. Congress has directed EPA to issue a progress report to the House and Senate Committees on Appropriations and relevant Congressional authorizing committees to describe EPA's implementation Of the NRC recommendations. This report provides Congress, stakeholders, and the public with an update on the IRIS Program and EPA's progress toward implementing the NRC recommendations and improving the Program.

II. INTRODUCTION

The U.S. Environmental Protection Agency's (EPA) Integrated Risk Information System (IRIS) Program provides health effects information on chemicals to which the public may be exposed from releases to air, water, and land and through the use and disposal of chemicals. IRIS human health assessments provide a critical part ofthe scientific foundation for decisions to protect public health across EPA's programs and regions under an array of environmental laws. IRIS assessments are not regulations, but they are critical to Agency decisions. IRIS assessments may include reference doses, reference concentrations, cancer slope factors, and inhalation unit risks that can be combined with exposure information to characterize the public health risks of chemical substances. This information is used by EPA, together with other considerations, to make risk management decisions, including regulations. IRIS also is a resource for risk assessors and environmental and health professionals in state and local governments and other countries.

IRIS assessments are scientific reports that provide information on a chemical's hazard as well as quantitative dose-response information. These are developed from the evaluation of research studies on health effects combined with judgments regarding issues such as appropriate study choice, characterization of effects, and uncertainty factors, among others. When combined with specific exposure information, EPA's program and regional offices and stakeholders can use IRIS assessments to help characterize the public health risks of chemical substances in various situations. During the risk management phase, other information may also be considered, such as statutory and legal requirements, cost/benefit information, technological feasibility, and economic factors. Figure 1 illustrates the risk assessment and risk management paradigm and where IRIS assessments contribute information.

Because of the critical importance of IRIS for the Agency and beyond, a strong, vital and scientifically sound Program is key to providing needed health risk information. Over the past two years, EPA has worked to strengthen and streamline the IRIS Program, improving transparency and increasing the number of final assessments added to the IRIS database. Continually improving IRIS is a priority for the Agency, and efforts are underway to further strengthen and streamline this important Program.

The purpose of this progress report is to provide Congress, stakeholders, and the public with a n updāte on the IRIS Program and EPA's progress toward implementing the NRC recommendations. This report provides an overview of improvements made to the IRIS Program over the past several years, with a n emphasis on the most recent improvements.

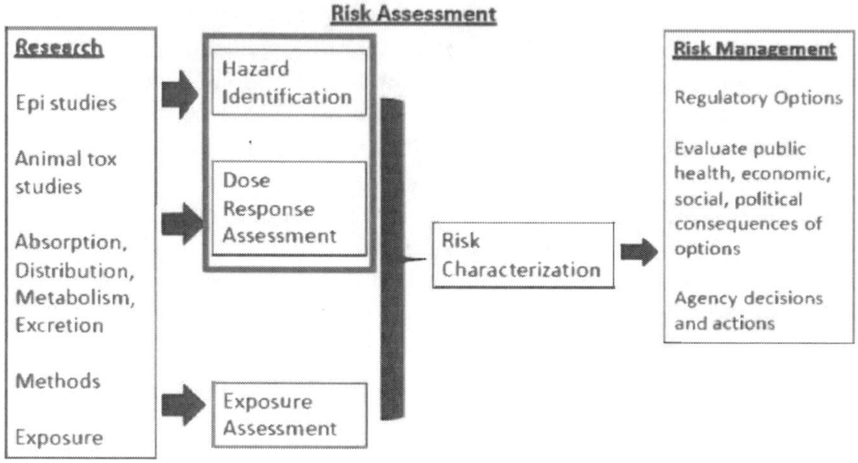

Adapted from the National Research Council rtsk assessment risk manag~ment paradigm (NRC 1983).

Figure 1. Risk Assessment Risk Management Paradigm (adapted from the National Research Council's paradigm, 1983). The red box shows the information included in IRIS assessments.

III. THE CONSOLIDATED APPROPRIATIONS ACT, 2012

On December 23, 2011, "The Consolidated Appropriations Act, 2012" was signed into law. The report language included direction to EPA for the IRIS Program related to recommendations provided by the National Research Council (NRC) in their review of EPA's draft IRIS assessment for formaldehyde. The NRC's recommendations, provided in Chapter 7 of their review report, offered suggestions for improving the development of IRIS assessments (see Tables 1 and 2). The report language directs EPA to incorporate, "as appropriate, based on chemical-specific datasets and biological effects," the NRC recommendations into the IRIS process. The report language also directs EPA to include documentation, in draft assessments released in fiscal year 2012, describing how the NRC recommendations were implemented. Further, the report language directs EPA to issue a progress report to the House and Senate Committees on Appropriations and relevant Congressional authorizing committees to describe implementation of the NRC recommendations for ongoing and new IRIS assessments.[1] The full report language is provided in Appendix 1.

IV. BACKGROUND

In May 2009, EPA Administrator Lisa Jackson announced a new IRIS assessment development process (Figure 2) designed to streamline, strengthen and improve transparency within the IRIS Program while ensuring the highest level of scientific quality and integrity and a renewed commitment to rigorous independent peer review.2

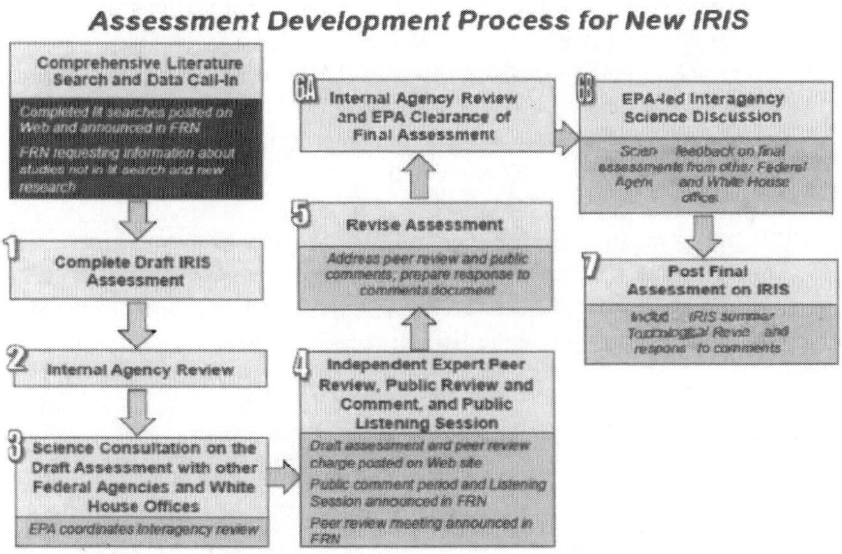

Figure 2. May 2OO9 IRIS Assessment Development Process.[3]

This new process was announced after a thorough review of the IRIS Program and in response to investigations by the Government Accountability Office (GAO) conducted in 2OO8. The GAO found that the IRIS database was at risk of becoming obsolete because EPA has not been able to routinely complete timely, credible assessments.[4] The May 2OO9 IRIS process included the following key features.[5,6]

- EPA would manage the IRIS Program and have final responsibility for the content of all IRIS assessments;
- The assessment development time was shortened to **23** months, a reduction of more than half the estimated time for an assessment to be developed under the previous process;
- The number of steps in the assessment development process was reduced from 14 to 7;
- Other federal agencies and White House offices would have the opportunity to provide scientific input at two points in the assessment development process, and the comments would be made publicly available;
- The assessment development process would include the opportunity for public comment and rely on an open, rigorous and independent external peer review;
- A public listening session would be offered for each chemical assessment; and
- Changes in EPA's scientific judgments during the process would be clearly documented and explained.

EPA has remained strongly committed to scientific integrity, public involvement, rigorous independent external peer review, and consultation with scientists at White House offices and other federal agencies. Opportunities for public involvement throughout the IRIS process include:

- Public nominations of substances to be considered for an assessment or reassessment through the IRIS Program.
- Public availability and opportunity to comment on a completed literature review for chemicals at the beginning of the assessment development process.
- Public availability of draft IRIS assessment documents and interagency comments for review and comment.
- Listening session, where the public can make comments or present information related to the draft assessment.
- Independent expert peer review meeting, which is open to the public and where members of the public may make formal comments and presentations.

V. OVERVIEW OF IMPROVEMENTS FROM MAY 2009 TO APRIL 2011

Since the new process was instituted in 2009, and as of the writing of this report, EPA has completed 24 assessments. In fiscal year 2009, EPA completed six final IRIS assessments after the new process was implemented. In fiscal year 2010, EPA completed 10 IRIS assessments and released nine for external peer review and public comment. In fiscal year 2011, EPA completed four assessments, including the long-awaited assessment for trichloroethylene (TCE). Thus far in fiscal year 2012, EPA has completed the final IRIS assessments for dichloromethane, tetrahydrofuran, tetrachloroethylene (perc) and the non-cancer assessment for dioxin. EPA remains strongly committed to continually improving the IRIS Program. From May 2009 to April 2011, EPA:

- Expanded the role of EPA's program and regional offices in nominating and prioritizing chemicals for assessment development to ensure that the IRIS Program is responsive to the most critical Agency needs;
- Hosted regular meetings between the IRIS Program and EPA's programs and regions to discuss individual IRIS assessments and the IRIS process;
- Created an IRIS logistics team to help create further efficiencies in the assessment development process; this team is charged with coordinating all administrative support, freeing up scientific staff to focus on the science of the assessments; and
- Developed the web-based Health and Environmental Research Online (HERO) database (www.epa.gov.hero), which promotes transparency by capturing and making available to the public the references and abstracts to scientific studies used in Agency health and environmental assessments.

VI. OVERVIEW OF EPA'S ROADMAP TO REVISIONS

In April 2O11, the National Research Council (NRC) released its "Review of the Environmental Protection Agency's Draft IRIS Assessment of Formaldehyde." In addition to offering comments specifically about EPA's draft formaldehyde assessment, the NRC included comments and recommendations to improve IRIS documents generally. A summary of the general recommendations is provided in Tables 1 (short-term) and 2 (long-term).

Table 1. National Research Council recommendations that EPA is implementing in the short-term[7]

General recommendations for completing the IRIS formaldehyde assessment that EPA will adopt for all IRIS assessments (p. 152)
1. To enhance the clarity of the document, the draft IRIS assessment needs rigorous editing to reduce the volume of text substantially and address redundancies and inconsistencies. Long descriptions of particular studies should be replaced with informative evidence tables. When study details are appropriate, they could be provided in appendices.
2. Chapter 1 needs to be expanded to describe more fully the methods of the assessment, including a description of search strategies used to identify studies with the exclusion and inclusion criteria articulated and a better description of the outcomes of the searches and clear descriptions of the weight-of-evidence approaches used for the various non-cancer outcomes. The committee emphasizes that it is not recommending the addition of long descriptions of EPA guidelines to the introduction, but rather clear concise statements of criteria used to exclude, include, and advance studies for derivation of the RfCs and unit risk estimates.
3. Standardized evidence tables for all health outcomes need to be developed. If there were appropriate tables, long text descriptions of studies could be moved to an appendix or deleted.
4. All critical studies need to be thoroughly evaluated with standardized approaches that are clearly formulated and based on the type of research, for example, observational epidemiologic or animal bioassays. The findings of the reviews might be presented in tables to ensure transparency.
5. The rationales for the selection of the studies that are advanced for consideration in calculating the RfCs and unit risks need to be expanded. All candidate RfCs should be evaluated together with the aid of graphic displays that incorporate selected information on attributes relevant to the database.
6. Strengthened, more integrative, and more transparent discussions of weight of evidence are needed. The discussions would benefit from more rigorous and systematic coverage of the various determinants of weight of evidence, such as consistency.

General Guidance for the Overall Process (p. 164)

7. Elaborate an overall, documented, and quality-controlled process for IRIS assessments.

8. Ensure standardization of review and evaluation approaches among contributors and teams of contributors; for example, include standard approaches for reviews of various types of studies to ensure uniformity.

9. Assess disciplinary structure of teams needed to conduct the assessments.

Evidence Identification: Literature Collection and Collation Phase (p. 164)

10. Select outcomes on the basis of available evidence and understanding of mode of action.

11. Establish standard protocols for evidence identification.

12. Develop a template for description of the search approach.

13. Use a database, such as the Health and Environmental Research Online (HERO) database, to capture study information and relevant quantitative data.

Evidence Evaluation: Hazard Identification and Dose-Response Modeling (p. 165)

14. Standardize the presentation of reviewed studies in tabular or graphic form to capture the key dimensions of study characteristics, weight of evidence, and utility as a basis for deriving reference values and unit risks.

15. Develop templates for evidence tables, forest plots, or other displays.

16. Establish protocols for review of major types of studies, such as epidemiologic and bioassay.

Selection of Studies for Derivation of Reference Values and Unit Risks (p. 165)

17. Establish clear guidelines for study selection.

18. Balance strengths and weaknesses.

19. Weigh human vs. experimental evidence.

20. Determine whether combining estimates among studies is warranted.

Calculation of Reference Values and Unit Risks (pp. 165-166)

21. Describe and justify assumptions and models used. This step includes review of dosimetry models and the implications of the models for uncertainty factors; determination of approp r iate points of departure (such as benchmark dose, no-observed-adverse-effect level, and lowest observed-adverse-effect level), and assessment of the analyses that underlie the points of departure.

22. Provide explanation of the risk-estimation modeling processes (for example, a statistical or biologic model fit to the data) that are used to develop a unit risk estimate.

23. Provide adequate documentation for conclusions and estimation of reference values and unit risks. As noted by the committee throughout the present report, sufficient support for conclusions in the formaldehyde draft IRIS assessment is often lacking. Given that the development of specific IRIS assessments and their conclusions are of interest to many stakeholders, it is important that they provide sufficient references and supporting documentation for their conclusions. Detailed appendixes, which might be made available only electronically, should be provided, when appropriate.

Table 2. National Research Council recommendations that EPA is implementing in the long-term[8]

Weight-of-Evidence Evaluation:
Synthesis of Evidence for Hazard Identification
(p. 165)

24. Review use of existing weight-of-evidence guidelines.

25. Standardize approach to using weight-of-evidence guidelines.

26. Conduct agency workshops on approaches to implementing weight-of-evidence .guidelines.

27. Develop uniform language to describe strength of evidence on noncancer effects.

28. Expand and harmonize the approach for characterizing uncertainty and variability.

29. To the extent possible, unify consideration of outcomes around common modes of action rather than considering multiple outcomes separately.

Calculation of Reference Values and Unit Risks (pp. 165-166)

30. Assess the sensitivity of derived estimates to model assumptions and end points selected. This step should include appropriate tabular and graphic displays to illustrate the range of the estimates and the effect of uncertainty factors on the estimates.

EPA agrees with the NRC's recommendations for the development of IRIS assessments and is fully implementing them consistent with the NRC's "Roadmap for Revision/' which viewed the full implementation of their recommendations as a multi-year process. In July 2011, EPA announced plans to further improve the IRIS Program, both as part of an ongoing effort to strengthen IRIS and also in response to the NRC recommendations.[9] At that time, EPA announced that it would implement the NRC recommendations using a phased approach. As such, EPA is making the most extensive changes to documents that are in the earlier steps of the assessment development process. For assessments that are in the later stages of development, EPA is implementing the recommendations, as feasible, without taking the assessments backwards to earlier steps of the process. Appendix 2 provides a figure illustrating the timing of implementing Phases 1, 2, and 3, and a description of these Phases is provided below.

Phase 1 of implementation focused on a subset of the short-term recommendations, such as editing and streamlining documents, increasing transparency and clarity, and using more tables, figures, and appendices to present information and data in assessments. Phase 1 also focused on assessments near the end of the development process and close to final posting. EPA is now in Phase 2 of implementation, which addresses all of the short-term recommendations from Table 1. EPA is implementing all of these recommendations but recognizes that achieving full and robust implementation of certain recommendations will be an evolving process with input and feedback from the public, stakeholders, and external peer review committees. EPA will soon publicly release two draft IRIS assessments that represent a major advancement in implementing the short-term NRC recommendations (recommendations from Table 1). Phase 3 of implementation will incorporate the longer-term recommendations made by the NRC (recommendations from Table 2).

This phased approach is consistent with the NRC's "Roadmap for Revision" as described in Chapter 7 of the formaldehyde review report. The NRC stated that "the committee recognizes that the changes suggested would involve a multi-year process and extensive effort by the staff at the National

Center for Environmental Assessment and input and review by the EPA Science Advisory Board and others."[10]

To respond to the NRC recommendations, and as part of Phase 2 of implementing them, draft IRIS assessments released in fiscal year 2O12 will be:

- Shorter, more concise and visual, with more graphs and tables used to represent data.
- Clearer and more transparent with respect to data, methods, and decision criteria used.
- More rigorously edited to eliminate inconsistencies and redundancies.

These assessments also will include a template for description of the literature search approach and will be linked to the Health and Environmental Research Online (HERO) database, to capture study information and relevant quantitative data.

Additionally, these assessments will:

- Evaluate and describe the strengths and weaknesses of critical studies in a more uniform way.
- Include toxicity values for multiple effects associated with the chemical, if applicable, and where there are suitable epidemiologic or experimental studies available.
- Take into account the Hill criteria in assessing evidence from epidemiological studies.
- Weigh the epidemiologic and experimental studies pertinent to each effect in a more consistent and integrative manner and make a clear statement about the strength of the evidence for the effect.
- Provide graphical presentations that synthesize related effects from multiple studies.
- Routinely consider the use of multiple data sets or combine multiple responses in deriving toxicity values, where appropriate.
- Indicate which criteria were most influential in selecting critical studies and toxicity endpoints.
- Describe and justify assumptions and models used to calculate toxicity values, with adequate documentation, including explanations ofthe risk estimation modeling processes.

Additionally, as part of Phase 2, EPA is evaluating existing guidelines to establish clearer criteria for study selection that will:

- Identify the strengths and weaknesses of studies for quantitative analysis.
- Describe criteria for selecting approaches to deriving toxicity values.
- Consider the possibility of combining multiple studies or effects for deriving toxicity values.

EPA also is developing graphic aids that will illustrate the range of concentrations evaluated in each study selected for quantitative assessment. This will help EPA to identify clusters of studies and low or high reference values that may be inconsistent with the larger body of literature.

Finally, EPA has created Chemical Assessment Support Teams to formalize an internal process to provide additional overall quality control for the development of IRIS assessments. This initiative uses a team approach to making timely, consistent decisions about the development of IRIS assessments across the Program. These teams will help ensure that the necessary disciplinary expertise is available when developing draft assessments, and they will provide a forum for identifying and addressing key issues prior to peer review. These teams also will help EPA objectively monitor progress in implementing the NRC recommendations on an assessment-by-assessment basis. As part of Phase 3, EPA is evaluating existing weight-of-evidence approaches and will adopt an approach or develop a new approach to consistently evaluate weight-of-evidence in IRIS assessments. EPA also will further work to develop systematic approaches to quantify uncertainty and variation. These actions respond to the long-term recommendations in Table 2.

The NRC recognized that a number of past IRIS assessments meet their expectations related to certain recommendations they provided. For example, in their report reviewing EPA's draft IRIS assessment for formaldehyde, the NRC endorsed an approach for presenting and deriving proposed RfC values that was used in the recent IRIS assessments for tetrachloroethylene and trichloroethylene, noting that "the approach described and illustrated in Figure 6-2 has been used in recent EPA assessments of tetrachloroethylene (EPA 2008b) and trichloroethylene (EPA 2009b)"[11]

Since April 2O11, EPA has been making strong progress toward implementing the NRC's recommendations to streamline assessments, making them more concise and readable, and using appendices. In a recent review of the IRIS Program, the Government Accountability Office (GAO) noted that, while challenges remained, they had reviewed two IRIS assessments that reflect changes made in response to the NRC suggestions. In reviewing the assessment of urea, which was finalized in July 2O11, GAO stated that "EPA streamlined the report by moving sections of text from the body to an appendix, which shortened the body ofthe assessment from 89 to **57** pages, making it more concise."

GAO also reviewed the draft IRIS assessment of diisobutyl phthalate (DIBP), which was undergoing Agency review (Step **2**) when EPA provided it to GAO, noting that it includes "(1) descriptive and pictorial explanations of the study selection methods used; (2) tables that, among other things, give side-by-side comparisons of studies considered in determining the oral reference dose for the chemical; and (3) brief descriptions of the strengths and weaknesses of various studies considered." GAO further noted that, for these two assessments, "it appears that EPA has begun to enhance the readability of its assessments by making changes that appear to be in line with the suggestions made by the National Academies."[12]

Figure 3 illustrates one aspect of EPA's efforts for Phase 1 of implementing the NRC recommendations. This figure provides assessment-specific information related to streamlining IRIS assessments by developing shorter, clearer and more concise documents with a reduced volume of text and a greater use of appendices for presenting certain types of information.

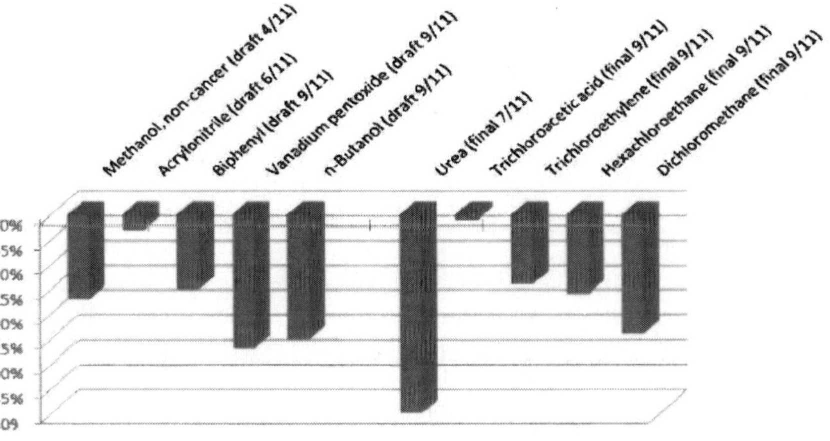

Figure 3. (Continued).

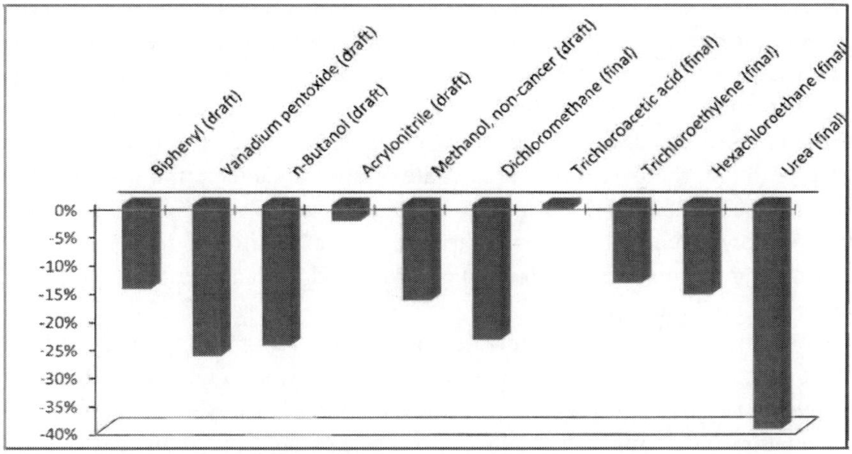

Figure 3. Percent change in number of pages in recent IRIS assessments (not including appendices) between prior and most recent version of assessments.

VII. REVISION INITIATIVES

In addition to the actions described above, EPA has developed several initiatives that are key aspects of EPA's plans to further improving the IRIS Program. Unless otherwise noted, these initiatives are part of EPA's response to the recommendations in Table 1.

EPA has presented these initiatives to various stakeholders at professional society meetings and has received positive feedback.

A. New Document Structure

As part of Phase 2 of implementing the NRC recommendations, EPA is improving the IRIS assessment template to substantially reduce the volume of text and address redundancies and inconsistencies in assessments. The new template provides a clear explanation of the literature review search strategy, study selection criteria, and methods used to develop the assessment. It also provides clear descriptions of the decision-making process that EPA uses in developing hazard identification and dose-response modeling.

The new document structure includes an Executive Summary in the beginning of each assessment that provides a concise summary of the major conclusions of the assessment related to hazard characterization and dose response analyses. Additionally, a newly developed Preamble describes the methods used to develop the assessment. Each assessment will include information on the literature search strategy used to develop the document, as well as the evaluation criteria and rationale used to make decisions about including and excluding studies in the assessment. Additional details about the' Preamble and literature search strategy are found in Sections VII B and C of this report.

As part of the new document structure, EPA is reorganizing the main body of the assessment into two sections, Hazard Identification and Dose-Response Analysis, to help further reduce the volume of text and redundancies. Information on chemical and physical properties and toxicokinetics is being moved to a supplemental materials appendix which will improve the flow of the document.

In the Hazard Identification chapter of the new document template, EPA will use evidence tables to present the key study findings that support how toxicological hazards are identified. Exposure-response arrays will be used as visual tools to inform the analysis. This chapter will provide a strengthened and more integrated and transparent discussion of the weight of the available evidence supporting the findings related to the chemical's hazard and toxicity. EPA also is developing standardized study summary tables, which will be included in the supplemental materials appendices, to present more detailed study summary information and key study characteristics.

EPA has used several of these approaches in the past in developing IRIS assessments and the IRIS Program will use these approaches consistently moving forward. For example, EPA's draft IRIS Toxicological Review of Biphenyl, which was released for public comment and external peer review in September 2O11, includes exposure-response arrays to provide a 'visual representation of points of departure for various effect endpoints resulting from exposure to the chemical. These arrays support EPA's decision-making in developing the toxicity analysis (e.g., identifying doses associated with specific effects; selecting critical studies and effects; etc.). Figure **4,** below, provides an example of an exposure-response array presented in the draft IRIS Toxicological Review of Biphenyl.[13]

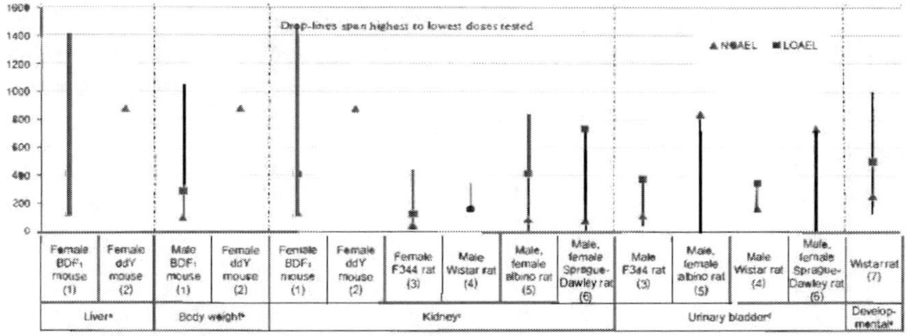

[a] Increased plasma liver enzymes in BDF_1 mice.

[b] Decreased body weight (>10% tower than controls) in BDF_1 mice.

[c] Increased incidences of kidney lesions including: mineralization in outer medulla in BDF_1 mice; renal pelvis transitional cell hyperplasia and hemosiderin deposits in F344 rats; kidney stone formation in Wistar rats; renal tubular atrophy in albino rats; renal tubular dilatation in Sprague-Dawley rats.

[d] Increased incidences of urinary bladder calculi or stones and hyperplasia in F344 rats and Wistar rats.

[e] I ncreased number of litters with fetal skeletal anomalies in Wistar rats.

(1) = Umeda et al., 2005; (2) = Imal et al., 1983; (3) = Umeda et al., 2002; (4) = Shiraiwa et al., 1989; (5) =Ambrose et al., 1960; (6) =Dow Chemical Co., 1953; (7) = Khera et al., 1979.

Figure 4. Exposure-response array, External Peer Review Draft IRIS Toxicological Review of Biphenyl.

The Dose-Response Analysis section of the new document structure provides a clear explanation of the rationale used to select and advance studies for consideration in calculating toxicity values. Key data supporting the dose-response analysis are reported and the methodology and derivation of toxicity values are described. In addition, details of the dose-response analysis, including the data, models, methods, and software, are provided in an appendix and described in sufficient detail to allow for independent replication and verification. The section also includes figures showing candidate reference values for comparison across studies and endpoints. Finally, the Dose-Response Analysis section of the new document structure includes clear documentation of the conclusions and estimation of the toxicity values.

EPA has used reference value arrays in the past and has received positive feedback from independent reviewers on this approach. These types of arrays will be used on a routine basis in the future where the data allow. For example, EPA's 2010 draft "Reanalysis of Key Issues Related to Dioxin Toxicity and Response to NAS Comments" (dioxin reanalysis) provided an array of candidate reference doses. This is illustrated in Figure **S.** The Science Advisory Board (SAB), in their review of the draft dioxin reanalysis, noted that "Figures **4.3** and 4.4 in the Report show quantitative comparisons across the RfDs and benchmark dose lower bounds calculated from the animal and epidemiological studies. These figures are useful in understanding the quantitative similarities ...in these calculations."[14]

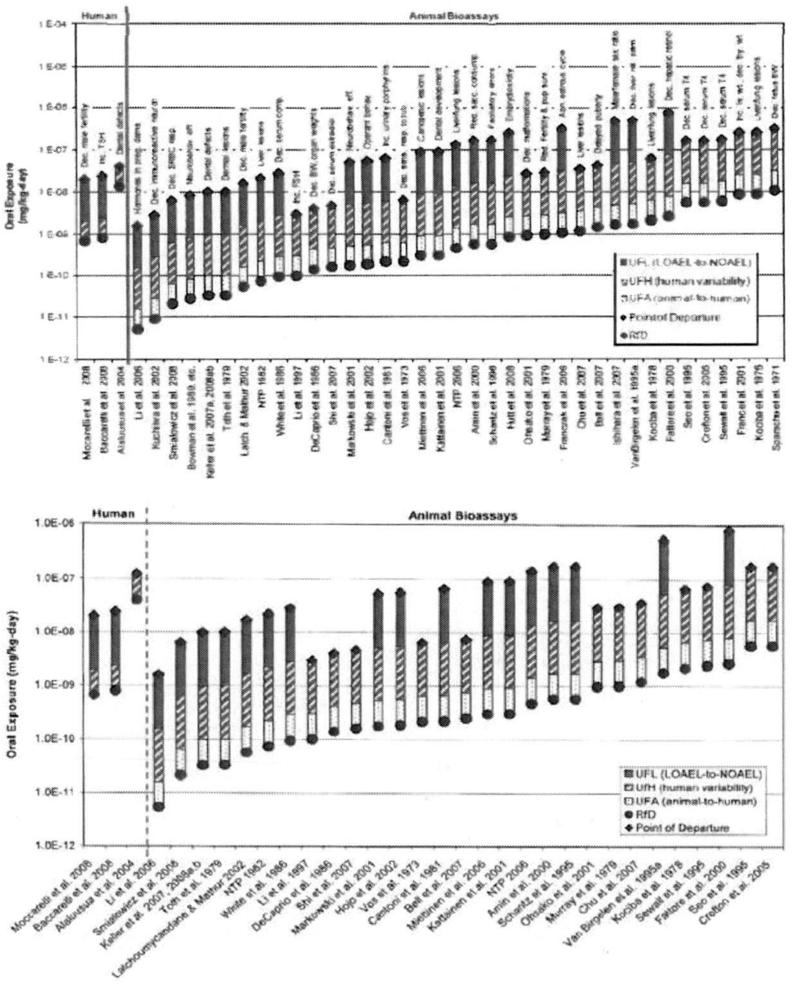

Figure 5. Reference value array from EPA's 2010 draft "Reanalysis of Key Issues Related to Dioxin Toxicity and Response to NAS Comments".

B. IRIS Assessment Preamble

In their review report, the NRC recommended that EPA expand Chapter 1 of IRIS assessments to "describe more fully the methods of the assessment, including a description of search strategies used to identify studies with the exclusion and inclusion criteria clearly articulated and a better description of the outcomes of the searches (a model for displaying the results of literature searches is provided later in this chapter) and clear descriptions of the weight-of-evidence approaches used for the various noncancer outcomes. The committee emphasizes that it is not recommending the addition of long descriptions of EPA guidelines to the introduction, but rather clear and concise statements of criteria used to exclude, include, and advance studies for derivation of the RfCs and unit risk estimates."[15]

In accordance with this NRC recommendation, and as part of Phase 2 of implementation, EPA is replacing Chapter 1 of IRIS assessments with a Preamble that will describe the application of existing EPA guidance and the methods and criteria used in developing the assessments. The term "Preamble" is used to emphasize that these methods and Criteria are being applied consistently across IRIS assessments. The new Preamble includes discussions about the following topics:

- Identifying and selecting pertinent studies
- Evaluating the quality of individual studies
- Weighing the overall evidence of each effect
- Selecting studies for derivation of toxicity values
- Deriving toxicity values

These topics correspond to the five steps that the NRC identified in their review report, "Elements of the key steps in the development of a draft IRIS assessment,"[16] which is shown in Figure 6 below.

For each topic, the Preamble summarizes and cites EPA guidance while also incorporating other NRC recommendations. The new Preamble, for example, emphasizes the disclosure of literature search strategies, standardized study evaluation criteria based on the type of study, the use of evidence tables to summarize study details, strengthened and more transparent discussions of weight-of-evidence, and graphical displays of candidate toxicity values.

The Preamble will be included in all new IRIS assessments as they are released for external peer review. The Preamble will be discussed at the first public meeting of the new SAB panel that will be dedicated to reviewing IRIS assessments (additional details are provided in Section VII F).

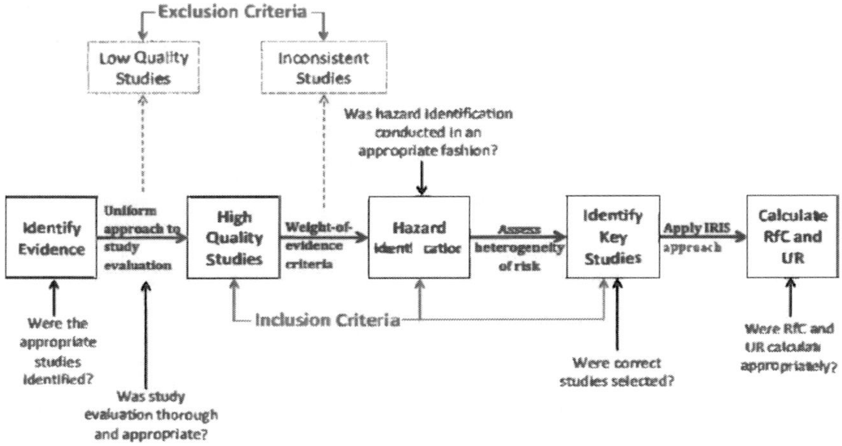

Figure 6. Elements of the key steps in the development of a draft IRIS assessment (Figure 7-2 of the NRC review report of formaldehyde).[17]

C. Literature Search Strategy

EPA's new document structure includes a detailed description of the literature search strategy and study evaluation process that EPA uses to develop IRIS assessments. This description, which is part of Phase 2 of EPA's implementation of the NRC recommendations, will be included in new IRIS assessments as they are developed. In discussing the literature search strategy in IRIS assessments, EPA will describe how the scientific literature was gathered and emphasize how studies were selected to be included in the document, and, if applicable, explain the rationale for excluding potentially relevant studies from the assessment. This section of the new document structure will be specific to each chemical assessment. It is designed to provide enough information that an independent literature search would be able to replicate the results of the literature search used by EPA in developing the assessment. In this section, EPA will provide a link to an external database (www.epa.gov.hero) that contains the references that were cited in the document, along with those that were considered for inclusion in the assessment, but not cited.

D. Weight of Evidence

EPA is currently developing a formal framework to establish conclusions about the weight of evidence for health effects other than cancer. In the meantime, the Agency is using existing guidelines that address these issues to inform its assessments. In addition, EPA is taking a more systematic approach to analyze the available human and animal toxicity data in IRIS assessments. This is being done as part of Phase 2 of EPA's implementation ofthe NRC recommendations.

In conducting this analysis and developing the synthesis, EPA evaluates the data for the:

- strength of the relationship between the exposure and response and the presence of a dose-response relationship;
- specificity of the response to chemical exposure and whether the exposure precedes the effect;
- consistency of the association between the chemical exposure and response; and
- biological plausibility of the response or effect and its relevance to humans.

EPA uses this weight of evidence approach to identify the potential hazards associated with chemical exposure.

EPA recognizes the benefit of adopting a formal weight-of-evidence framework that includes standardized classification of causality. As part of Phase 3 of implementing the NRC recommendations, the Agency is planning and will convene a workshop on adapting weight-of-evidence procedures for effects other than cancer. The NRC discussed current approaches to hazard identification, which involves answering the question: Does the agent cause the adverse effect? [pp 159-16O).[18] The NRC recommended that the answer to this question should be a clear statement on the strength of the evidence of causation, following a standardized classification to avoid ambiguity and to ensure comparability among different agents and outcomes. The NRC offered as examples the Surgeon General's 2OO4 report on smoking and the monographs of the International Agency for Research on Cancer and the U.S. National Toxicology Program.

During 2O12, EPA will convene a workshop to discuss approaches to weight-of-evidence characterization. EPA will identify the various approaches that are currently in use and compare their strengths and limitations. The workshop will include scientists with expertise in the classification of chemicals for various health effects. The workshop will be open to the public, with opportunity for written and oral comments. EPA is currently discussing the workshop within the Agency and considering how it will be conducted. EPA will publicly announce details about the workshop in a Federal Register Notice and on the IRIS website (www.epa.gov.iris).

EPA will consider the comments and discussions from the workshop to either adopt or adapt existing approaches to weight-of-evidence classification for use in subsequent IRIS assessments. These activities respond to recommendations 24-29 in Table 2.

E. Early Peer Consultation

In addition to the public listening session and public comment period that are already part of the IRIS assessment development process, EPA will increase the use of public peer consultation workshops to enhance the input of the scientific community as assessments are designed. These workshops will be open to the public with opportunity for oral and written comments. Information regarding specific peer consultation workshops will be announced to the public in advance of the meetings through a Federal Register Notice and on the IRIS website (www.epa.gov.iris). While not an explicitly stated recommendation of the NRC in the formaldehyde review report, early peer consultation workshops fit with the spirit of the NRC's overall group of recommendations. The goal of these workshops will vary. For example, the workshops may focus on the state-of-the-science for a particular chemical or provide a forum for discussion with experts about certain cross-cutting scientific issues that may impact the development of a scientifically complex assessment. The first of these peer consultation workshops will focus on mouse lung tumors as they relate to human cancer risk. This is an important issue for the IRIS assessments for naphthalene, styrene, and ethylbenzene. This workshop will be open to the public, with opportunity for oral and written comments. It is anticipated that the workshop will convene in summer 2012.

EPA has occasionally held these types of workshops in the past and the Agency plans to incorporate early peer consultation more frequently in the future. For example, in December 2010, EPA convened a peer consultation workshop, which included a panel of external independent experts, to discuss approaches to assessing the cumulative risks of mixtures of phthalates. The experts were asked to review recommendations provided by the NRC in their

2008 report, "Phthalates and Cumulative Risk: The Tasks Ahead," for evaluating the cumulative risk of mixtures of phthalates and to propose any additional methods or approaches, not captured in the NRC report, that could be appropriate to use for this type of cumulative assessment. The public was invited to register and attend the workshop as observers and give brief oral comments at the conclusion of each workshop day. Additionally, the public was invited to provide written comments regarding the subject matter under discussion. EPA is using the comments and recommendations resulting from the workshop, including the public comments, to develop a draft cumulative assessment for six phthalates. EPA convened a similar peer consultation workshop for dioxin in 2OO9, hosting a public, three-day, scientific workshop to discuss key issues related to dioxin's toxicity. The results of this workshop informed EPA's development of EPA's 2O1O draft "Reanalysis of Key Issues Related to Dioxin Toxicity and Response to NAS Comments."

F. Dedicated Chemical Assessment Advisory Committee

While the NRC focused their recommendations exclusively on the development of draft IRIS assessments, and did not specifically address peer review, EPA is seeking to make some changes in this area to further improve IRIS. Specifically, EPA has established a dedicated Chemical Assessment Advisory Committee, under the auspices of EPA's Science Advisory Board (SAB), to provide advice to EPA on draft IRIS Toxicological Reviews and the IRIS Program. I n November 2O11, EPA announced a request for the public to nominate potential panelists for a new, standing advisory panel including experts in the fields of public health, epidemiology, toxicology, modeling, biostatistics, and risk assessment.[19] The SAB will take public comment on the proposed nominated panel before assembling the final Chemical Assessment Advisory Committee. EPA will send the first assessment to the new committee for a consultation as soon as it is established, followed by a face-to-face meeting in summer 2012. Two additional meetings with the panel are anticipated for later i n the year. EPA will consult the panel for peer review of a range of IRIS assessments and seek advice on how the IRIS Program implements the NRC recommendations. A schedule for meetings and draft assessments to be reviewed is under development and will be announced to the public in a Federal Register Notice and on the IRIS website (www.epa.gov.iris). The first assessments that will be sent to the new SAB Chemical Assessment Advisory Committee for review will be the draft IRIS assessments for ammonia and two trimethylbenzenes (1,2,4- and 1,3,5-). These draft assessments, which will be i n the new document structure, will be the first released for public comment and external peer review that were developed during Phase 2 of implementation. These assessments represent a major adyancement for the IRIS Program in implementing the NRC recommendations.

CONCLUSION

EPA's IRIS Program proyides critically important health effects information on chemicals to which the public may be exposed from releases to air, water, and land and through the use and disposal of chemicals. IRIS assessments are critical to risk assessors and decision-makers at EPA and beyond. EPA's Board of Scientific Counselors (BOSC) has reyiewed the IRIS Program on seyeral occasions. They haye recognized the importance and quality of the IRIS Program, noting that the IRIS website receiyes about **8** million yisits annually, "a testament to the yalue of IRIS as a resource" and stating that "the comprehensiyeness, transparency, and consistency ofthe IRIS approach haye made it into the internationally recognized standard in hazard characterization."[20]

EPA recognizes and embraces this role, and understands that a strong, scientifically rigorous IRIS Program is of critical importance. Oyer the past two years, EPA has worked to strengthen and streamline the IRIS Program, improying transparency and increasing the number of final assessments added to the IRIS database while maintaining a firm commitment to public engagement and rigorous expert peer reyiew. Continually improying the IRIS Program is a priority for the Agency. EPA will continue to pursue excellence in the IRIS Program, using the best science with integrity and efficiency in support of EPA's mission to protect the health of the American public.

APPENDIX 1. LANGUAGE FROM THE "CONSOLIDATED APPROPRIATIONS ACT, 2012" REPORT LANGUAGE [21]

Integrated Risk Information System (IRIS) - In lieu of the directives contained in H. Rept. 112-151 regarding the Integrated Risk Information System, the conferees agree to the following:

(1) Fundamental improvements to the policies and practices of this program are necessary to ensure that IRIS assessment reflect the highest standard of scientific integrity.

(2) The Agency shall incorporate, as appropriate, based on chemical-specific datasets and biological effects, the recom mendations of Chapter 7 of the National Research Council's Review of the Environmental Protection Agency's Draft IRIS Assessment of Formaldehyde into the IRIS process.

(3) The Agency shall issue a progress report to House and Senate Committees on Appropriations and relevant Congressional authorizing committees no later than March 1, 2012, describing its implementation of the National Research Council's Chapter 7 recommendations for ongoing and new assessments.

(4) For draft assessments released in fiscal year 2012, the Agency shall include documentation describing how the Chapter 7 recommendations of the National Academy of Science (NAS) have been implemented or addressed, including an explanation for why certain recommendations were not incorporated.

(5) The Agency shall contract with NAS to conduct up to three reviews of IRIS assessments that EPA seeks to make final. Reviews shall include an evaluation of whether the recommendations it made in previous reviews, including in Chapter 7 of the National Research Council's Review of the Environmental Protection Agency's Draft IRIS Assessment of Formaldehyde, have been implemented. Reviews are not intended to unduly delay the Agency's risk assessment process. The conferees further direct NAS to complete any reviews authorized by this paragraph by no later than 18 months after the date that EPA and the NAS have agreed to the terms of the review. One of these NAS reviews shall be a study of the cancer and non-cancer hazards from oral exposure to inorganic arsenic. The NAS review of inorganic arsenic shall incorporate the direction provided in House Report 112-151 regarding par ameters of the study. Additional reviews will be chosen by NAS from a representational sample of IRIS assessments and NAS will notify Congress directly of these choices.

(6) Further, the conferees strongly believe any current and future IRIS assessments must not only be grounded in sound, objective, and peer-reviewed science and methodologies but should also provide risk managers with realistic values that will result in enhanced protection of human health.

APPENDIX 2. TIMELINE FOR PHASES 1, 2, AND 3 OF IMPLEMENTING THE NRC RECOMMENDATIONS FOR IRIS

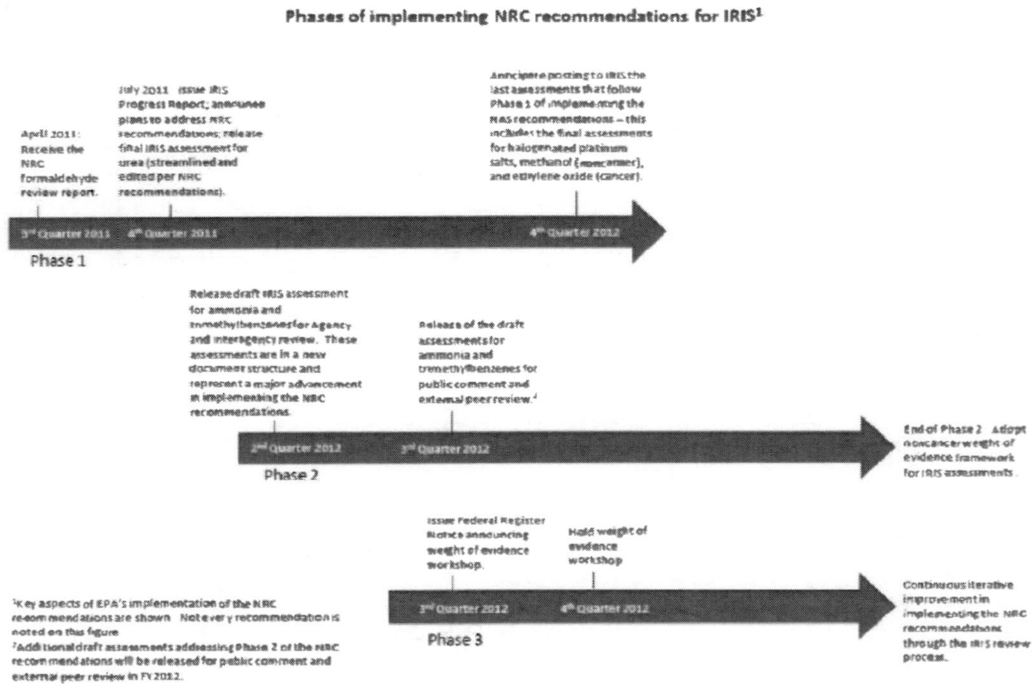

End Notes

[1] Pub. L. No. 112-74, Consolidated Appropriations Act, 2012

[2] http://yosemite.epa.gov/opa/admpress.nsf/48f0fa7dd51f9e9885257359003f5342/065e2c61afea0917852575bd0064c9db!OpenDocument

[3] http://www.epa.gov/iris/pdfs/IRIS_PROCESS_FLOW_CHART.PDF

[4] Government Accountability Office, 2008. *Chemical Assessments-Low Productivity and New Interagency Review Process Limit the Usefulness an Credibility of EPA's Integrated Risk Information System.*

[5] IRIS Progress Report, July 2011. http://www.epa.gov/iris/pdfs/irisprogressreport2011.pdf

[6] http://www.epa.gov/iris/pdfs/IRIS_PROCESS_MEMO.5.21.09.PDF

[7] National Research Council, 2011. Review of the Environmental Protection Agency's Draft IRIS Assessment of Formaldehyde.

[8] National Research Council, 2011. Review of the Environmental Protection Agency's Draft IRIS Assessment o Formaldehyde.

[9] http://yosemite.epa.gov/opa/admpress.nsf/d0cf6618525a9efb85257359003fb69d/a3fcd6083819706785257 8cb00666c4d!OpenDocumen

[10] National Research Council, 2011. Review of the Environmental Protection Agency's Draft IRIS Assessment of Formaldehyde.

[11] National Research Council, 2011. Review of the Environmental Protection Agency's Draft IRIS Assessment of Formaldehyde.

[12] Government Accountability Office, 2011. Chemical Assessments: Challenges Remain with EPA's Integrated Risk Information System Program.

[13] US Environmental Protection Agency. Draft Toxicological Review of Biphenyl. EPA/635/R-11/005A. September 2011.

[14] U.S. Environmental Protection Agency Science Advisory Board, EPA-SAB-011-014. SAB Review of EPA's Reanalysis of Key Issues Related to Dioxin Toxicity and Response to NAS Comments (May 2010). August 2011.

[15] National Research Council, 2011. Review of the Environmental Protection Agency's Draft IRIS Assessment of Formaldehy

[16] National Research Council, 2011. Review of the Environmental Protection Agency's Draft IRIS Assessment of Formaldehyde.

[17] National Research Council, 2011. Review of the Environmental Protection Agency's Draft IRIS Assessment of Formaldehyde.

[18] National Research Council, 2011. Review of the Environmental Protection Agency's Draft IRIS Assessment of Formaldeh

[19] http://www.gpo.gov/fdsys/pkg/FR-2011-11-18/pdf/2011-29916.pdf

[20] Board of Scientific Counselors Human Health Risk Assessment Research Program Review, April 2008 http://www.epa.gov/osp/bosc/pdf/hhra0804rpt.pdf

[21] Pub. L. No. 112-74, Consolidated Appropriations Act, 2012

In: Health Assessments …
Editor: Marcus Rowe

ISBN: 978-1-63321-166-7

Chapter 3

THE TOXIC SUBSTANCES CONTROL ACT (TSCA): A SUMMARY OF THE ACT AND ITS MAJOR REQUIREMENTS*

Linda-Jo Schierow

SUMMARY

This report summarizes the Toxic Substances Control Act (TSCA) and the major regulatory programs dealing with chemical production and distribution in U.S. commerce.

The President's Council on Environmental Quality proposed comprehensive federal legislation in 1971 to identify and control potentially dangerous chemicals in U.S. commerce that were not adequately regulated under other environmental statutes. President Ford signed TSCA into law on October 11, 1976. Subsequently, five titles were added to address specific concerns—asbestos in 1986 (Title II, P.L. 99-519), radon in 1988 (Title III, P.L. 100-551), lead in 1992 (Title IV, P.L. 102-550), environmental and energy issues in schools in 2007 (Title V, P.L. 110-140), and formaldehyde emissions from composite wood products in 2010 (Title VI, P.L. 111-199). In 2008, Congress added provisions to Title I, Section 6 and Section 12, banning certain activities with respect to elemental mercury (P.L. 110-414).

---

* This is an edited, reformatted and augmented version of a Congressional Research Service publication, CRS Report for Congress RL31905, prepared for Members and Committees of Congress, from www.crs.gov, dated April 1, 2013.

TSCA authorizes EPA to identify potentially dangerous chemicals in U.S. commerce that should be subject to federal control. The act authorizes EPA to gather and disseminate information about production, use, and possible adverse effects to human health and the environment of existing chemicals, and to issue "test rules" that require manufacturers and processors of potentially dangerous chemicals to conduct and report the results of scientific studies to fill information gaps. For chemicals new to U.S. commerce, TSCA requires pre-market screening and regulatory tracking of new chemical products.

If EPA identifies unreasonable risks associated with existing or new chemicals, TSCA requires the agency to initiate rulemaking to reduce risks to a reasonable level. EPA may regulate the manufacture, importation, processing, distribution, use, and/or disposal of chemicals. TSCA provides a variety of regulatory tools to EPA, ranging in severity from a total ban on production, import, and use to a requirement that a product must bear a warning label at the point of sale. However, TSCA directs EPA to use the least burdensome option that can reduce risk to a level that is reasonable, given the benefits provided by the chemical product or process.

Title I of the original statute establishes the core program, directs EPA to control risks from polychlorinated biphenyls (PCBs), and bans certain activities with respect to elemental mercury. Title II directs EPA to set standards for asbestos mitigation in schools and requires asbestos contractors to be trained and certified. Title III directs EPA to provide technical assistance to states that choose to support radon monitoring and control. Title IV provides similar assistance with respect to abatement of lead-based paint hazards. Title V addresses environmental issues at schools, including energy efficiency. Finally, Title VI directs EPA to set standards for emissions of formaldehyde from composite wood products.

INTRODUCTION

The Environmental Protection Agency (EPA) is responsible for identifying and regulating toxic substances in U.S. commerce under the authority of the Toxic Substances Control Act (TSCA). This report defines key terms, provides a brief history of toxic substances control law, and describes key provisions of TSCA. In addition, this report lists several references for more detailed information about the act and provides a table that cross references sections of the *U.S. Code* with corresponding sections of the act. The report is descriptive rather than analytic, highlights key provisions rather than providing a comprehensive inventory of the act's numerous sections, and addresses authorities and limitations imposed by statute, rather than the status of EPA implementation or other policy issues.

OVERVIEW

The Toxic Substances Control Act (15 U.S.C. 2601 *et seq.*) authorizes the EPA to screen existing and new chemicals used in manufacturing and commerce to identify potentially dangerous products or uses that should be subject to federal control. Both naturally occurring and synthetic chemicals are subject to TSCA, with the exception of chemicals regulated under other federal laws concerning food, drugs, cosmetics, firearms, ammunition, pesticides, tobacco, or mixtures. EPA may require manufacturers and processors of chemicals to conduct and report the results of tests to determine the effects of potentially dangerous chemicals on living things. Based on test results and other information, EPA must regulate the manufacture, importation, processing, distribution, use, and/or disposal of any chemical that presents an unreasonable risk of injury to human health or the environment. A variety of regulatory tools is available to EPA under TSCA, ranging in severity from a total ban on production, import, and use to a requirement that a product bears a warning label at the point of sale. TSCA directs EPA to use the least burdensome option that can reduce risk to a level that is reasonable, given the benefits provided by the chemical product or process.

Table 1. Toxic Substances Control Act and Major Amendments (codified as 15 U.S.C. 2601-2671)

Year	Act	Public Law Number
1976	Toxic Substances Control Act	P.L. 94-469
1986	Asbestos Hazard Emergency Response Act	P.L. 99-519
1988	Radon Program Development Act	P.L. 100-551
1990	Radon Measurement	P.L. 101-508, §10202
1990	Asbestos School Hazard Abatement Reauthorization Act	P.L. 101-637
1992	Residential Lead-Based Paint Hazard Reduction Act of 1992	P.L. 102-550
2007	Energy Independence and Security Act of 2007, subtitle E - Healthy High-Performance Schools	P.L. 110-140
2008	Mercury Export Ban Act of 2008	P.L. 110-414
2010	Formaldehyde Standards for Composite Wood Products Act	P.L. 111-199

Federal legislation to control toxic substances was originally proposed in 1971 by the President's Council on Environmental Quality. Its report, "Toxic Substances," defined a need for comprehensive legislation to identify and control chemicals whose manufacture, processing, distribution, use, and/or disposal was potentially dangerous and not adequately regulated under other environmental statutes. The House and Senate each passed bills in both the 92nd and 93rd Congresses (in 1972 and 1973, respectively), but controversies over the scope of chemical screening prior to commercial production and distribution, level of costs, and the relationship to other regulatory laws stalled final action. Episodes of environmental contamination—including contamination of the Hudson River and other waterways by PCBs, the threat of stratospheric ozone depletion from chlorofluorocarbon (CFC) emissions, and contamination of agricultural produce by polybrominated biphenyls (PBBs) in the state of Michigan—together with more exact estimates of the costs of imposing toxic substances controls, opened the way for final passage of the legislation. President Ford signed the TSCA into law on October 11, 1976. The original legislation included a single title, which has since been designated Title I.

TSCA (Title I) directs EPA to

- require manufacturers and processors to conduct tests for existing chemicals if (1) their manufacture, distribution, processing, use, or disposal may present an unreasonable risk of injury to health or the environment; or they are to be produced in substantial quantities and the potential for environmental release or human exposure is substantial or significant; (2) existing data are insufficient to predict the effects of human exposure and environmental releases; and (3) testing is necessary to develop such data (§4);
- prevent future risks through pre-manufacture screening and regulatory tracking of new chemical products (§5);
- control unreasonable risks already known, including risks from polychlorinated biphenyls (PCBs), as well as risks for existing chemicals that may be discovered in the future (§6); and
- gather and disseminate information about chemical production, use, and possible adverse effects to human health and the environment (§8).

Authorization for appropriations for these activities and a state grant program for control of toxic substances in the environment expired on

September 30, 1983, although appropriations for these programs have continued.

In October 2008, Congress amended TSCA Title I when it enacted the Mercury Export Ban Act of 2008 (P.L. 110-414). It prohibits certain activities with respect to elemental mercury.

Subsequently, five titles have been added to TSCA to address specific concerns—asbestos in 1986 (Title II, P.L. 99-519), radon in 1988 (Title III, P.L. 100-551), lead in 1992 (Title IV, P.L. 102-550), schools in 2007 (Title V, P.L. 110-140), and formaldehyde in 2010 (Title VI, P.L. 111- 199). Title II directs EPA to set standards for asbestos mitigation in schools, and requires asbestos contractors to be trained and certified. Title III directs EPA to provide technical assistance to states that choose to support radon monitoring and control. Title IV provides similar assistance with respect to abatement of lead-based paint hazards. Title V addresses environmental issues at schools, including energy efficiency. Finally, Title VI establishes limits on emissions of formaldehyde from composite wood products.

TITLE I

Testing of Chemicals

Many chemicals, even some in widespread use, are not well characterized in terms of their potential health and environmental effects. One of the major goals of TSCA was to induce the development of test data by producers (i.e., manufacturers, importers, and processors) of chemicals in commerce. Section 4 of TSCA directs EPA to require the development of test data on existing chemicals when certain conditions prevail: (1) the manufacture, processing, distribution, use, or disposal of the chemical "may present an unreasonable risk;" or (2) the chemical is produced in very large volume and there is a potential for a substantial quantity to be released into the environment or for substantial or significant human exposure. Under either condition, EPA must issue a rule requiring tests if (a) existing data are insufficient to resolve the question of safety, and (b) testing is necessary to develop the data.

Because there were more than 55,000 chemicals in U.S. commerce at the time EPA was to begin developing test rules, Congress established a special interagency committee to help EPA determine which chemicals should be considered first, and to coordinate testing needs and efforts among government agencies. At least every six months the Interagency Testing Committee (ITC) must consider candidate chemicals for inclusion on a list of substances that the ITC recommends to EPA for development and promulgation of test rules. TSCA directs the ITC to "designate" a subset of chemicals on the list for EPA action within 12 months. The list can contain no more than 50 "designated" chemicals at any time. When a chemical is designated, EPA has one year to respond by issuing a proposed test rule or a notice explaining why no testing is needed.

TSCA requires the ITC to consider the following factors when it makes listing decisions: (1) quantity of the substance to be manufactured, (2) quantity of the chemical in environmental releases, (3) number of people who will be exposed occupationally and the duration of exposure, (4) extent of non-occupational human exposure, (5) similarity of the chemical to any other chemical known to present an unreasonable risk, (6) existence of data concerning environmental or health effects of the chemical, (7) the quantity of information to be gained by testing, and (8) the availability of facilities and personnel for performing testing. Chemicals known or suspected to cause or contribute to cancer, gene mutations, or birth defects are to be assigned a higher priority. In response to information that indicates "there may be a reasonable basis to conclude that a chemical ... presents or will present a significant risk of serious or widespread harm to human beings from cancer, gene mutations, or birth defects," TSCA requires EPA action to prevent or reduce that risk or publication of a finding that the risk is not unreasonable.

Pre-manufacture Notification for New Chemicals or Uses

TSCA (§5) requires manufacturers, importers, and processors to notify EPA at least 90 days prior to producing or otherwise introducing a new chemical product into the United States. Any information or test data that is known to, reasonably ascertainable by, or in possession of the notifier, and that might be useful to EPA in evaluating the chemical's potential adverse effects on human health or the environment, must be submitted to EPA at the same time. TSCA also requires EPA to be notified when there are plans to produce, process, or use an existing chemical in a way that differs from previously permitted uses, if the Administrator has determined by rule that new uses of the chemical may produce significant changes in human and environmental exposures and therefore require notification. The 90-day notice provides EPA with the opportunity to evaluate the chemical use and, if necessary, to prohibit or limit such activity before it occurs to prevent unreasonable risk of injury to human health or the environment.

EPA has 45 days after notification (or up to 90 days if it extends the period for good cause) to evaluate the potential risk posed by the chemical. If EPA determines that there is a reasonable basis to conclude that the substance presents or will present an unreasonable risk, the Administrator must promulgate requirements to protect adequately against such risk. Alternatively, EPA may determine that the proposed activity related to a chemical does not present an unreasonable risk; this decision may be based on the available data, or, when no data exist to document the effects of exposure, on what is known about the effects of chemicals in commerce with similar chemical structures and used in similar ways.

The purpose of EPA's screening procedure is to identify potential hazards, and control them before use of a chemical becomes widespread. If data are inadequate to make an informed judgment and (1) manufacture, processing, distribution in commerce, use, or disposal may present an unreasonable risk; or (2) a chemical is to be produced in substantial quantities, and the potential for environmental release or human exposure is substantial or significant, EPA may issue a proposed order to prohibit or limit such activities until sufficient data are submitted.

Although the legislative history of TSCA includes a presumption that testing of new products would take place before they were widely used, either as the chemical was developed, or as its markets grew, TSCA also forbids promulgation of blanket testing requirements for all new chemicals. This reflects concern that uniform testing requirements might stifle innovation in the chemical industry. Thus, EPA must decide which chemicals, or which categories of chemicals, warrant the costs of pre-market testing. EPA reviews approximately 1,000 new chemical manufacturing notices annually.

Regulatory Controls for Hazardous Chemicals

TSCA requires EPA to regulate manufacturing, processing, distribution in commerce, use, or disposal of a chemical if it will present an unreasonable risk of injury to health or the environment, and the risk cannot be reduced to a sufficient degree under another federal law administered by EPA. The alternative means available to EPA for controlling chemical hazards that present unreasonable risks are specified in Section 6 of TSCA. EPA has the authority to

- prohibit or limit the amount of production or distribution of a substance in commerce;
- prohibit or limit the production or distribution of a substance for a particular use;
- limit the volume or concentration of the chemical produced;
- prohibit or regulate the manner or method of commercial use;
- require warning labels and/or instructions on containers or products;
- require notification of the risk of injury to distributors and, to the extent possible, consumers;
- require record-keeping by producers;
- specify disposal methods; and
- require replacement or repurchase of products already distributed.

EPA also may impose any of these requirements in combination or for a specific geographical region. However, EPA is required by TSCA to regulate only "to the extent necessary to protect adequately" against a risk, and to use the "least burdensome" regulatory approach, even in controlling unreasonable risks.

Two chemical substances are directly addressed in Title I: PCBs and elemental mercury. TSCA directs EPA to regulate PCBs and to ban most uses. In addition, TSCA prohibits the sale, distribution, or transfer of elemental mercury by federal agencies.

Information Gathering

Section 8 of TSCA requires EPA to develop and maintain an inventory of all chemicals, or categories of chemicals, manufactured or processed in the United States. The first version of this inventory identified approximately 55,000 chemicals in commerce in 1979. All chemicals not on the inventory are, by definition, "new" and subject to the notification provisions of Section 5. These chemicals must be added to the inventory if they enter U.S. commerce. Chemicals need not be listed if they are only produced in very small quantities for purposes of experimentation or research.

To aid EPA in its duties under TSCA, the agency was granted considerable authority to collect information from industries. EPA may require maintenance of records and reporting of chemical identities, names, and molecular structures; categories of use; amounts manufactured and processed for each category of use; descriptions of byproducts resulting from manufacture, processing, use, and disposal; environmental and health effects; number of individuals exposed; number of employees exposed and the duration of exposure; and manner or method of chemical disposal.

Manufacturers, processors, and distributors of chemicals are required to maintain records of significant adverse reactions to health or the environment alleged to have been caused by a substance or mixture. Records of adverse effects on the health of employees must be retained for 30 years from the date of reporting. Industry also must submit lists and copies of health and safety studies. Studies showing adverse effects previously unknown must be submitted to EPA as soon as they are completed or discovered.

Imminent Hazards

Section 7 provides EPA authority to take emergency action through the district courts to control a chemical substance or mixture which presents an imminent and unreasonable risk of serious widespread injury to health or the environment.

Relation to Other Laws

Section 9 requires EPA to report cases of chemical risk to other federal agencies with the authority to prevent or reduce the risk. If the other agency determines that the activity does not present the risk described by EPA in its report, or initiates action under the law to protect against the risk, TSCA prohibits EPA from taking regulatory action with respect to such risk. For statutes under EPA's jurisdiction, TSCA gives the Administrator discretion to decide if a risk can best be handled under the authority of TSCA.

Enforcement and Judicial Review

Section 11 authorizes EPA to inspect any facilities subject to TSCA requirements and to issue subpoenas requiring attendance and testimony of witnesses, production of reports and documents, answers to questions and other necessary information. Section 13 mandates TSCA enforcement at the national borders by the Treasury Department.

Section 15 identifies acts prohibited under TSCA, while Section 16 describes penalties for acts violating these prohibitions, as well as recourse available to anyone accused of such violations. Section 16 authorizes civil penalties, not to exceed $25,000 per violation per day, and affords the defendant an opportunity to request a hearing before an order is issued and to petition for judicial review of an order after it is issued. Criminal penalties also are authorized for willful violations. Section 17 provides jurisdiction to U.S. district courts in civil actions to enforce TSCA Section 15 by restraining or compelling actions that violate or comply with it, respectively. Chemicals may be seized and condemned if their manufacture, processing, or distribution violated the act. Section 19 authorizes any person to file a petition for judicial review of specified rules within 60 days of issuance under TSCA. The court is directed to set aside specified rules if they are not supported by substantial evidence in the rulemaking record taken as a whole.

Section 20 authorizes civil suits by any person against any person in violation of the act. It also authorizes suits against EPA to compel performance of nondiscretionary actions under TSCA. Section 21 provides the public with the right to petition for the issuance, amendment, or repeal of a rule requiring toxicity testing of a chemical, regulation of the chemical, or reporting.

Confidential Business Information

Section 14 provides broad protection of proprietary confidential information about chemicals in commerce. Disclosure by EPA employees of such information generally is not permitted, except to other federal employees, or when necessary to protect health or the environment. Data from health and safety studies of chemicals are not protected unless their disclosure would reveal a chemical process or chemical proportion in a mixture. Wrongful disclosure of confidential data by federal employees is prohibited, and may result in criminal penalties.

Chemical Categories

Section 26 allows EPA to impose regulatory controls on categories of chemicals, rather than on a case-by-case basis. However, EPA cannot regulate a group merely because it is composed of new chemical substances.

State Preemption

TSCA Section 18 preempts state actions that establish or continue in effect requirements applicable to a chemical substance or mixture that is regulated under TSCA Section 5 or 6, unless the state requirement is identical to the federal requirement, implements another federal law, or prohibits use of the substance or mixture within the state. However, a state may ask EPA to allow a state requirement that provides a significantly higher degree of protection from risk than does the federal requirement.

Other Provisions

TSCA Section 10 directs EPA to conduct and coordinate among federal agencies research, development, and monitoring that is necessary to the purposes of the act.

Section 12 excludes chemical products manufactured for export from TSCA requirements except for reporting and record keeping requirements in Section 8. In 2008, Congress excluded elemental mercury from this exemption, banning its export beginning in 2013, with the exception of mercury contained in coal. Other exemptions for essential uses may be granted by rule.

Section 22 waives compliance when in the interest of national defense.

Section 23 provides protection of employees who assist in carrying out the provisions of the act (i.e., “whistle-blowers”).

The potential effects of TSCA rules on employment must be monitored by EPA, according to Section 24.

Section 25 mandates study of the need for indemnification of people affected by federal laws administered by EPA and of the feasibility of establishing a standard classification system for chemical substances and of storing and retrieving information about them.

Section 26 authorizes data sharing and cooperative action to facilitate TSCA implementation between EPA and other federal agencies. It also authorizes collection of fees for EPA processing of data submitted in response to an order under Section 4 or 5. EPA is directed to establish an office to assist the regulated community. The agency also must establish a procedure to ensure disclosure of financial interests in the regulated community by EPA employees. Final orders issued under TSCA must contain a statement of basis and purpose. Finally, Section 26 established within EPA a new Assistant Administrator for Toxic Substances.

TSCA Section 27 authorizes research and development of test methods for chemicals by the Public Health Service in cooperation with EPA.

Grants to states are authorized by Section 28 to establish and operate programs to prevent or eliminate unreasonable risks to health or the environment.

Section 29 authorized appropriations through 1983. An annual report is mandated by Section 30.

TITLE II (ASBESTOS IN BUILDINGS)

Growing public concern about the presence of potentially hazardous asbestos in buildings, especially in schools, led to congressional efforts to address this problem. Title II of TSCA, the Asbestos Hazard Emergency Response Act (AHERA), was enacted in 1986 (P.L. 99-519) and amended in July 1988 (P.L. 100-368). It required EPA to set standards by October 1987, for responding to the presence of asbestos in schools. The standards, set at levels adequate to protect public health and the environment, identify appropriate response actions that depend on the physical condition of asbestos. Schools, in turn, were required to inspect for asbestos-containing material, and to develop and implement a plan for managing any such material. Plans for managing asbestos were to be submitted by schools before May 1989, and implementation was to begin by July 1989. The law contains no deadlines for schools to complete implementation.

Title II requires asbestos contractors and analytical laboratories to be certified, and schools to use certified persons for abatement work. Training and accreditation requirements also apply to inspectors, contractors, and workers performing asbestos abatement work in all public and commercial buildings. EPA may award training grants to nonprofit organizations for asbestos health and safety programs. However, authorization of appropriations for this grant program expired September 30, 1995. Other Title II requirements (such as mandates that buildings be inspected for asbestos) have not been extended to non-school buildings.

To enforce requirements, TSCA authorizes EPA to take emergency action with respect to schools if school officials do not act to protect children. The act also authorizes citizen action with respect to asbestos-containing material in a school and to compel action by EPA, either through administrative petition or judicial action. Civil penalties not to exceed $5,000 are authorized for violations such as failing to conduct an inspection or to develop a school management plan.

Concern about how schools would pay for required actions was addressed in separate legislation (the Asbestos School Hazard Abatement Act of 1984, or ASHAA, P.L. 98-377). It established a program offering grants and interest-free loans to schools with serious asbestos problems and demonstrated financial need. Although EPA for several years did not request funding for this program, Congress appropriated funds. Authorization of appropriations for this program expired September 30, 1995, and Congress has not appropriated funds since FY1993; a total of $382 million in grant and loan funds were appropriated from FY1984 through FY1993. Repaid ASHAA loans are returned to an Asbestos Trust Fund, established in TSCA Title II, to become a dedicated source of revenues for future asbestos control projects.

TITLE III (RADON PROGRAMS)

In October 1988, Congress amended TSCA by adding Title III—Indoor Radon Abatement (15 U.S.C. 2661 *et seq.*, P.L. 100-551). The basic purpose of Title III is to provide financial and technical assistance to the states that choose to support radon monitoring and control; neither monitoring nor abatement of radon is required by the act.

Title III required EPA to update its pamphlet "A Citizen's Guide to Radon," to develop model construction standards and techniques for controlling radon levels within new buildings, and to provide technical assistance to states. EPA is to provide technical assistance by establishing an information clearinghouse; publishing public information materials; establishing a national database of radon levels detected, organized by state; providing information to professional organizations representing private firms involved in building design and construction; submitting to Congress a plan for providing financial and technical assistance to states; operating cooperative projects with states; conducting research to develop, test, and evaluate radon measurement methods and protocols; developing and demonstrating new methods of radon measurement and mitigation, including methods that are suitable for use in nonresidential child care facilities; operating a voluntary program to rate radon measurement and mitigation devices and methods and the effectiveness of private firms and individuals offering radon-related services; and designing and implementing training seminars. The proficiency rating program and certification for training programs collect fees for service, and therefore are meant to be self-supporting, but Congress authorized $1.5 million to be appropriated to establish these programs. Congress authorized $3 million to be appropriated for each of three years beginning in 1989 for the other provisions of Sections 303, 304, and 305.

A matching grant program was established for the purpose of assisting states in developing and implementing programs for radon assessment and mitigation. For this program, $30 million was authorized to be appropriated over three years, with funds targeted to states or projects that made efforts to ensure adoption of EPA's model construction standards and techniques for new buildings; gave preference to low-income persons; or addressed serious and extensive radon contamination problems or had the potential to reduce risk or to develop innovative assessment techniques, mitigation measures, or management approaches.

Other sections of Title III require EPA to conduct a study to determine the extent of radon contamination in schools; identify and list areas of the United States with a high probability of having high levels of indoor radon; make grants or cooperative agreements to establish and operate at least three regional radon training centers; and provide guidance to federal agencies on radon measurement, risk assessment, and remedial measures.

All authorizations for appropriations specific to this title expired September 30, 1991, although appropriations have continued.

TITLE IV (LEAD EXPOSURE REDUCTION)

The 102nd Congress added Title IV to TSCA when it enacted the Residential Lead-Based Paint Hazard Reduction Act of 1992 as Title X in the Housing and Community Development Act of 1992 (P.L. 102-550). TSCA Title IV aims to accelerate federal efforts to reduce risks to young children who daily are exposed to lead-based paint in their homes. In addition, it was intended to stimulate development of lead inspection and hazard abatement services in the private sector, while ensuring that the services provided and any products employed are reliable and effective in reducing risk. To these ends, Title IV directs EPA to

- promulgate definitions of lead-contaminated dust, lead-contaminated soil, and lead-based paint hazards;

- ensure that people engaged in detection and control of lead hazards are properly trained and that contractors are certified;
- publish requirements for the accreditation of training programs for workers;
- develop criteria to evaluate the effectiveness of commercial products used to detect or reduce risks associated with lead-based paint;
- establish protocols, criteria, and minimum performance standards for laboratory analysis of lead in paint films, soil, and dust;
- establish a program to certify laboratories as qualified to test substances for lead content; and
- publish and distribute to the public a list of certified or accredited environmental sampling laboratories.

Title IV explicitly applies these requirements to federal facilities and activities that may create a lead hazard.

In addition, Congress directed EPA to conduct a study of lead hazards due to renovation and remodeling activities that may incidentally disturb lead-based paint. EPA is required to promulgate guidelines for the renovation and remodeling of buildings or other structures when these activities might create a hazard.

Title IV directs EPA to establish a clearinghouse and hotline to distribute information about the hazards of lead-based paint, how to avoid exposure and reduce risk, and new technologies for removing or immobilizing lead-based paint. In addition, Congress mandated development of a lead hazard information pamphlet; public education and outreach activities for health professionals, the general public, homeowners, landlords, tenants, consumers of home improvement products, the residential real estate industry, and the home renovation industry; and information to be distributed by retailers of home improvement products to provide consumers with practical information related to the hazards of renovation where lead-based paint may be present.

Title IV authorizes states to propose programs to train and certify inspectors and contractors engaged in the detection or control of lead-based paint hazards. States also may develop the required informational pamphlets. TSCA requires EPA to promulgate a model state program that may be adopted by any state. Congress gave EPA the authority to approve or disapprove authorization for state proposals and to provide grants for states to develop and implement authorized programs. A federal program must be established, administered, and enforced by EPA in each state without an authorized program.

The Department of Health and Human Services also has responsibilities under Title IV of TSCA. It mandates a study by the Centers for Disease Prevention and Control (CDC) and the National Institute for Environmental Health Sciences to determine the sources of lead exposure to children who have elevated lead levels in their bodies. The National Institute for Occupational Safety and Health is directed to study ways of reducing occupational exposure to lead during abatement activities.

The act established a rule-making docket to ensure the availability to the general public of all documents submitted to agencies that are relevant to regulatory decisions pursuant to this legislation. The docket is required to include the drafts of all proposed rules submitted by EPA to the President's Office of Management and Budget (OMB), written comments on the drafts, and written responses to comments. In addition, the agency must provide an explanation for any major change to a proposed rule that appears in the final rule, and such changes may not be made based on information not filed in the docket. Dockets are required to be established in each EPA regional office.

Congress authorized to be appropriated "such sums as may be necessary" for TSCA Title IV.

In addition to amending TSCA, Title X of the Housing and Community Development Act of 1992 authorized grants to states for risk assessments and lead-based paint removal and immobilization in private housing for low-income residents; establishing state training, certification, or accreditation programs for inspectors and abatement contractors; and research at the Department of Housing and Urban Development (HUD). Authorization for appropriations for these grants expired September 30, 1994, but appropriations have continued. Title X directed HUD to establish guidelines for federally supported work involving risk assessments, inspections, interim controls, and abatement of lead-based paint hazards. In addition, the National Institute for Occupational Safety and Health (NIOSH) was provided $10 million for training people who remove or immobilize paint.

TITLE V (REDUCING RISKS IN SCHOOLS)

At the end of 2007, the 110th Congress added a fifth title to TSCA, subtitled Healthy High-Performance Schools. Enacted as Title IV, Subtitle E (§461) of P.L. 110-140, the Energy Independence and Security Act of 2007, TSCA Title V authorizes EPA to establish a state grant program to provide technical assistance for EPA programs to schools and develop and implement state school environmental health programs. State programs must include standards for school building design, construction, and renovation, and identify ongoing school building environmental problems and recommended solutions. Environmental problems specifically mentioned in the law include "contaminants, hazardous substances, and pollutant emissions." EPA's authority to provide grants expires five years after the date of enactment.

Title V requires the EPA Administrator, in consultation with the Secretary of Education and the Secretary of Health and Human Services, to issue voluntary guidelines within 18 months of Title V enactment for selecting sites for schools (presumably new schools). The guidelines are to account for the "special vulnerability of children to hazardous substances or pollution exposures in any case in which the potential for contamination at a potential school site exists," modes of transportation available to students and staff, efficient use of energy, and potential use of a school at the site as an emergency shelter.

Title V also requires the EPA Administrator, in consultation with the Secretary of Education and the Secretary of Health and Human Services, to issue voluntary guidelines within two years of enactment for developing and implementing state environmental health programs for schools. These guidelines must take into account the findings of federal initiatives established under "relevant federal law with respect to school facilities," including initiatives related to water and energy conservation authorized by Sections 431 through 441, and work related to high-performance green buildings authorized by Section 492 of P.L. 110-140. In particular, the guidelines must take into account "environmental problems, contaminants, hazardous substances, and pollutant emissions"; natural day lighting; ventilation; heating and cooling; moisture control and mold; maintenance, cleaning, and pest control; acoustics; and "other issues relating to the health, comfort, productivity, and performance of occupants of the school facilities." In addition, Title V requires that the guidelines provide "technical assistance on siting, design, management, and operation of school facilities"; collaborate with children's environmental health centers in school environmental investigations"; assist states and the public to better understand and improve the environmental health of children; and take into account "the special vulnerability of children in low-income and minority communities to exposures from contaminants, hazardous substances, and pollutant emissions."

Several provisions in Title V refer to entities established under other sections of the Energy Independence and Security Act of 2007 (P.L. 110-140). For example, Title V contains directives for the Federal Director of the Office of Federal High-Performance Green Buildings in the General Services Administration, which was created by Section 436(a). In addition, there is reference to the national high-performance green building clearinghouse established in Section 423(1) "to carry out public outreach to inform individuals and entities of the information and services [related to high-performance green buildings] available governmentwide." Title V requires the Federal Director to ensure, "to the maximum extent practicable," that the public clearinghouse "receives and makes available information on the exposure of children to environmental hazards in school facilities." The EPA Administrator is directed to prepare an annual report to Congress on activities carried out under Title V authority, and this report also must be made available to the public through the clearinghouse.

For the purposes of carrying out the provisions of Title V, Congress authorized appropriations of $1.5 million annually through 2013.

TITLE VI (LIMITING FORMALDEHYDE EMISSIONS)

In July 2010, Congress enacted the Formaldehyde Standards for Composite Wood Products Act (P.L. 111-199), adding a new Title VI to TSCA. The new title mandates specific formaldehyde emission standards for hardwood plywood, medium-density fiberboard, and particleboard that is sold, supplied, offered for sale, or manufactured in the United States. The standards are phased in over two years from enactment and are based on the voluntary national formaldehyde emissions standards established by ASTM International (formerly known as the American Society for Testing and Materials), method ASTM E-1333-96 (2002).

The standards apply to plywood, particleboard, and medium-density fiberboard in the form of an unfinished panel or incorporated into a finished good. Certain products are excluded, including many forms of lumber and panels used for outdoor applications, such as structural plywood, prefabricated wood I-joists, most windows, antiques or other previously owned goods, and composite wood products used inside automobiles, trucks, rail cars, boats, and aircraft.

EPA is required to promulgate regulations ensuring compliance with the emission standards and must include provisions relating to labeling, chain of custody requirements, sell-through provisions; ultra low-emitting formaldehyde resins, finished goods, third-party testing and certification; auditing and reporting of third-party certifiers; recordkeeping; enforcement, laminated products; and exceptions for products and components containing "de minimis amounts" of composite wood products. The new law prohibits stockpiling of products manufactured before the effective date of the act for sale after that date. Also prohibited is any requirement for labeling products manufactured prior to the "designated date of manufacture."

P.L. 111-199 requires an annual report to Congress on the status of implementation and the extent to which relevant industries have achieved compliance. Finally, the act directs the Secretary of Housing and Urban Development to update regulations concerning formaldehyde emissions from composite wood in manufactured homes (24 Code of Federal Regulations 3280.308) to ensure that the standards established by TSCA Title VI are implemented.

Table 2. Major U.S. Code Sections, Toxic Substances Control Act (codified as 15 U.S.C. 2601-2692)

15 U.S.C.	Section Title	Toxic Substances Control Act (as amended)
Subchapter I	Control of Toxic Substances	
2601	Findings, policy and intent	§2
2602	Definitions	§3
2603	Testing of chemical substances and mixtures	§4
2604	Manufacturing and processing notices	§5
2605	Regulation of hazardous chemical substances and mixtures	§6
2606	Imminent hazards	§7
2607	Reporting and retention of information	§8
2608	Relationship to other federal laws	§9
2609	Research, development, collection, dissemination, and utilization of data	§10
2610	Inspections and subpoenas	§11
2611	Exports	§12
2612	Entry into customs territory of the United States	§13
2613	Disclosure of data	§14
2614	Prohibited acts	§15
2615	Penalties	§16
2616	Specific enforcement and seizure	§17
2617	Preemption	§18
2618	Judicial	§19
2619	Citizens' civil actions	§20
2620	Citizens' petitions	§21

Table 2. (Continued)

15 U.S.C.	Section Title	Toxic Substances Control Act (as amended)
2621	National defense waiver	§22
2622	Employee protection	§23
2623	Employment effects	§24
2624	Studies	§25
2625	Administration	§26
2627	Development and evaluation of test methods	§27
2628	Authorization of appropriations	§28
2629	Annual report	§29
Subchapter II	Asbestos Hazard Emergency Response	
2641	Congressional findings and purpose	§201
2642	Definitions	§202
2643	EPA regulations	§203
2644	Requirements if EPA fails to promulgate regulations	§204
2645	Submission to state Governor	§205
2646	Contractor and laboratory accreditation	§206
2647	Enforcement	§207
2648	Emergency authority	§208
2649	State and federal law	§209
2650	Asbestos contractors and local educational agencies	§210
2651	Public protection	§211
2652	Asbestos ombudsman	§212
2653	EPA study of asbestos-containing material in public buildings	§213
2654	Transition rules	§214
2655	Worker protection	§215
Subchapter III	Indoor Radon Abatement	
2661	National goal	§301
2662	Definitions	§302
2663	EPA's citizen guide	§303
2664	Model construction standards and techniques	§304
2665	Technical assistance to states for radon programs	§305
2666	Grant Assistance to states for radon programs	§306
2667	Radon in schools	§307
2668	Regional radon training centers	§308
2669	Study of radon in federal buildings	§309
2670	Regulations	§310
2671	Additional authorizations	§311
Subchapter IV	Lead Exposure Reduction	

15 U.S.C.	Section Title	Toxic Substances Control Act (as amended)
2681	Definitions	§401
2682	Lead-based paint activities training and certification	§402
2683	Identification of dangerous levels of lead	§403
2684	Authorized state programs	§404
2685	Lead abatement and measurement	§405
2686	Lead hazard information pamphlet	§406
2687	Regulations	§407
2688	Control of lead-based paint at federal facilities	§408
2689	Prohibited acts	§409
2690	Relationship to other federal law	§410
2691	General provisions relating to administrative proceedings	§411
2692	Authorization of appropriations	§412
Subchapter V	Healthy High-Performance Schools	
2695	Grants for healthy school environments	§501
2695a	Model guidelines for siting of school facilities	§502
2695b	Public outreach	§503
2695c	Environmental health program	§504
2695d	Authorization of appropriations	§505
Subchapter VI	Formaldehyde Standards for Composite Wood Products	
2697	Formaldehyde Standards	§601

Note: This table shows only the major code sections. For more detail and to determine when a section was added, the reader should consult the official printed version of the U.S. Code.

SELECTED REFERENCES

Bergeson, Lynn L. TSCA: Toxic Substances Control Act. Basic Practice Series. Chicago, IL: Section of Environment, Energy, and Resources, *American Bar Association*, 2000. 151 p.

Bergeson, Lynn L., Lisa M. Campbell, and Lisa Rothenberg. TSCA and the Future of Chemical Regulation, *EPA Administrative Law Reporter*, 2000, v. 15, n. 4, 23 p.

CRS Report RS21688, Lead-Based Paint Poisoning Prevention: Summary of Federal Mandates and Financial Assistance for Reducing Hazards in Housing, by Linda-Jo Schierow.

In: Health Assessments …
Editor: Marcus Rowe

ISBN: 978-1-63321-166-7

Chapter 4

INTEGRATED RISK INFORMATION SYSTEM (IRIS) FREQUENT QUESTIONS*

Environmental Protection Agency

- What is IRIS?
- What substances are in IRIS?
- What is an RfD and RfC?
- What is a cancer weight-of-evidence descriptor?
- What is a cancer slope factor and unit risk?
- How does EPA decide which substances to add or update?
- How do I generate a Multiple Substances Report?
- I am interested in Inhalation Toxicology values….should I search using the term "air" or "inhalation"?
- What is the process for developing IRIS assessments?
- How are IRIS toxicity values used?
- How do I contact the IRIS Hotline?
- What is the role of IRIS assessments in risk assessment and risk management?

* This is an edited, reformatted and augmented version of the following EPA website: (http://www.epa.gov/IRIS/help_ques.htm#whatiris) viewed April 2014.

WHAT IS IRIS?

EPA's Integrated Risk Information System (IRIS) is a human health assessment program that evaluates risk information on effects that may result from exposure to environmental contaminants. Through the IRIS Program, EPA provides the highest quality science-based human health assessments to support the Agency's regulatory activities. The IRIS database contains information for more than 540 chemical substances containing information on human health effects that may result from exposure to various substances in the environment. IRIS is prepared and maintained by the EPA's National Center for Environmental Assessment (NCEA) within the Office of Research and Development (ORD).

The heart of the IRIS system is its collection of searchable documents that describe the health effects of individual substances and that contain descriptive and quantitative information in the following categories:

- *Noncancer effects:* Oral reference doses and inhalation reference concentrations (RfDs and RfCs (http://www.epa.gov/iris/help_ gloss. htm#r), respectively) for effects known or assumed to be produced through a nonlinear (possibly threshold) mode of action. In most instances, RfDs and RfCs are developed for the noncarcinogenic effects of substances.
- *Cancer effects:* Descriptors that characterize the weight of evidence for human carcinogenicity, oral slope factors, and oral and inhalation unit risks for carcinogenic effects. Where a nonlinear mode of action is established, RfD and RfC values may be used.

WHAT SUBSTANCES ARE IN IRIS?

A complete alphabetical list of the substances in IRIS is available at the A to Z List of IRIS Substances (http://cfpub.epa.gov/ncea/iris/index.cfm? fuseaction=iris.showSubstanceList). Use Search the IRIS Database (http:// www.epa.gov/IRIS/search_keyword.htm) by substance name or CASRN to search for IRIS assessments for a specific substance. Other specific search criteria are available as well. You can also search multiple substances at once using Compare IRIS Values (http://cfpub.epa.gov/ncea/ iris/compare.cfm).

WHAT IS AN RFD AND RFC?

EPA's IRIS is a human health assessment program that evaluates quantitative and qualitative risk information on effects that may result from exposure to specific chemical substances found in the environment. The IRIS database contains information that can be used to support the first two steps (hazard identification (http://www.epa.gov/risk/hazardous-identification.htm) and dose-response evaluation (http://www.epa.gov/risk/dose-response.htm)) of the risk assessment process (http://www.epa.gov/risk/health-risk.htm). When supported by available data, IRIS provides oral reference doses (RfDs) and inhalation reference concentrations (RfCs) for chronic non-cancer health effects, and oral slope factors and inhalation unit risks for carcinogenic effects. Combined with specific exposure information, government and private entities use IRIS to help characterize public health risks of chemical substances in a site-specific situation and thereby support risk management decisions designed to protect public health.

More specifically, the reference dose (RfD) and reference concentration (RfC) provide quantitative information for use in risk assessments for health effects known or assumed to be produced through a nonlinear (possibly threshold) mode of action. The RfD (expressed in units of mg of substance/kg body weight-day) is defined as an estimate (with uncertainty spanning perhaps an order of magnitude) of a daily exposure to the human population (including sensitive subgroups) that is likely to be without an appreciable risk of deleterious effects during a lifetime. An RfD can be derived from a no-observed-adverse-effect level (NOAEL), lowest-observed-adverse-effect level (LOAEL), or benchmark dose, with uncertainty factors generally applied to reflect limitations of the data used. The inhalation RfC (expressed in units of mg of substance/m^3 air) is analogous to the oral RfD but provides a continuous inhalation exposure estimate. The inhalation RfC considers toxic effects for both the respiratory system (portal of entry) and effects peripheral to the respiratory system (extrarespiratory or systemic effects). Reference values may also be derived for acute ($\leq$24 hours), short-term (>24 hours, up to 30 days), and subchronic (>30 days, up to approximately 10% of the life span) exposure durations, all of which are derived based on an assumption of continuous exposure throughout the duration specified. RfDs and RfCs are generally used in noncancer health assessments.

What Is a Cancer Weight-of-Evidence Descriptor?

A cancer weight-of-evidence (WOE) (http://www.epa.gov/iris/help_gloss.htm#w) descriptor is used by IRIS to describe a substance's potential to cause cancer in humans and the conditions under which the carcinogenic effects may be expressed. This judgment is independent of consideration of the agent's carcinogenic potency. Under EPA's 1986 guidelines for carcinogen risk assessment, the WOE was described by categories "A through E"—Group A for known human carcinogens through Group E for agents with evidence of noncarcinogenicity. Under the EPA's 2005 guidelines for carcinogen risk assessment, a narrative approach, rather than categories, is used to characterize carcinogenicity. Five standard weight-of-evidence descriptors (*Carcinogenic to Humans, Likely to Be Carcinogenic to Humans, Suggestive Evidence of Carcinogenic Potential, Inadequate Information to Assess Carcinogenic Potential, and Not Likely to Be Carcinogenic to Humans*) are used as part of the narrative.

What Is a Cancer Slope Factor and Unit Risk?

Cancer slope factors and unit risks are used to estimate the risk of cancer associated with exposure to a carcinogenic or potentially carcinogenic substance. A slope factor is an upper bound, approximating a 95% confidence limit, on the increased cancer risk from a lifetime exposure to an agent by ingestion. This estimate, usually expressed in units of proportion (of a population) affected per mg of substance/kg body weight-day, is generally reserved for use in the low-dose region of the dose-response relationship, that is, for exposures corresponding to risks less than 1 in 100. A unit risk is an upper-bound excess lifetime cancer risk estimated to result from continuous exposure to an agent at a concentration of 1 μg/L in water or 1 $\mu g/m^3$ in air. The interpretation of unit risk for a substance in drinking water would be as follows: if unit risk = 2×10^{-6} per μg/L, 2 excess cancer cases (upper bound estimate) are expected to develop per 1,000,000 people if exposed daily for a lifetime to 1 μg of the substance in 1 liter of drinking water.

How Does EPA Decide Which Substances to Add or Update?

EPA develops a list of substances for IRIS assessment development on an annual basis. The IRIS program submits queries to EPA Program Offices and Regions and the public for nominations for new assessments or updates of assessments currently on IRIS. Substances are selected based on one or more of the following factors: (1) potential public health impact; (2) EPA statutory, regulatory, or program-specific implementation needs; (3) availability of new scientific information or methodology that might significantly change the current IRIS information; (4) interest to other governmental agencies or the public; and (5) availability of other scientific assessment documents that could serve as a basis for an IRIS assessment. The decision to assess any given chemical substance depends on available Agency resources. Availability of risk assessment guidance, guidelines, and science policy decisions may also have an impact on the timing of EPA's decision to assess a chemical substance.

The list of new or updated assessments is published in the Federal Register (FR) as part of the IRIS annual agenda.

How Do I generate a Multiple Substances Report?

To compare toxicity values for multiple substances, click on Compare IRIS Values (http://cfpub.epa.gov/ncea/iris/compare.cfm). This link allows you to generate summary reports of the toxicity values for multiple substances.

I am interested in Inhalation Toxicology values....should I search using the term "air" or "inhalation"?

Either search term is helpful. Older IRIS assessments generally use the term "air". This terminology has been updated and changed to "inhalation". We recommend that you search using both terms, perhaps individually.

What Is the Process for Developing IRIS Assessments?

EPA's process for developing IRIS assessments consists of: (1) a Federal Register announcement of EPA's IRIS agenda and call for scientific information from the public on the selected substances, (2) a search of the current scientific literature, a Federal Register announcement that the literature search is available on the IRIS internet site, and a call to submit additional scientific information on the substance, (3) development of a draft Toxicological Review or other assessment document , (4) internal peer consultation, (5) internal Agency Review, (6) Science Consultation with other Federal agencies and White House offices, (7) external peer review and public comment, (8) final internal Agency Review, Interagency Science Discussion and ORD management approval, and (9) posting on the IRIS database.

This process is described more fully via the IRIS Process (http://www.epa.gov/IRIS/process.htm) page.

HOW ARE IRIS TOXICITY VALUES USED?

IRIS provides hazard identification and dose-response assessment information. The information in IRIS can be used in combination with exposure information to characterize the public health risks of a given substance in a given situation. These risk characterizations can form the basis for risk-based decision-making, regulatory activities, and other risk management decisions designed to characterize and protect public health.

HOW DO I CONTACT THE IRIS HOTLINE?

Contact the IRIS Hotline via the Contact us (http://www.epa.gov/IRIS/contact_hotline.htm) page or at (202) 566-1676 (phone), (202) 566-1749 (fax), or hotline.iris@epa.gov (email).

WHAT IS THE ROLE OF IRIS ASSESSMENTS IN RISK ASSESSMENT AND RISK MANAGEMENT?

Risk assessment is a process that has been defined as "the characterization of the potential adverse health effects of human exposures to environmental hazards" (NRC, 1983). Estimates of environmental exposure are combined with the known adverse effects of exposure to determine an overall estimate of the potential public health risk.

A complete risk assessment consists of the following four steps:

1. Hazard identification;
2. Dose-response assessment;
3. Exposure assessment; and
4. Risk characterization.

Hazard identification involves the determination of whether exposure to an agent can cause an increased incidence of an adverse health effect, such as cancer or birth defects, and characterization of the nature and strength of the evidence of causation (NRC, 1994).

Dose-response assessment is the characterization of the relationship between exposure or dose and the incidence and severity of the adverse health effect. It includes consideration of factors that influence dose-response relationships such as intensity and patterns of exposure and age and lifestyle variables that could affect susceptibility. It can involve extrapolation of high-dose responses to low-dose responses and from animal responses to human responses (NRC, 1994).

Exposure assessment is the determination of the intensity, frequency, and duration of actual or hypothetical exposure of humans to the agent in question (NRC, 1994).

The risk characterization integrates information from the preceding components of the risk assessment and synthesizes an overall conclusion about risk that is complete, informative, and useful for decision makers (U.S. EPA, 2000). The risk characterization also addresses the uncertainty, assumptions, and scientific judgments of the previous three steps.

An IRIS health assessment consists of the hazard identification and dose-response assessment steps. The information in the IRIS assessment is combined with site- or problem-specific exposure assessments to provide the scientific support for EPA risk management decisions.

EPA considers risk assessment information along with social and economic factors, public health impacts, and statutes and regulations, in deciding how best to protect public health and the environment. Examples of risk management actions include deciding how much of a substance a company may discharge into a river; deciding which substances may be stored at a hazardous waste disposal facility; deciding to what extent a hazardous waste site must be cleaned up; setting permit levels for discharge, storage, or transport; establishing levels for air emissions; and determining allowable levels of contamination in drinking water.

REFERENCES AND FURTHER READING

NRC (National Research Council). 1983. Risk Assessment in the Federal Government: Managing the Process. National Academy Press, Washington, DC.

NRC. (1994) Science and Judgment in Risk Assessment. Committee on Risk Assessment of Hazardous Air Pollutants, Board on Environmental Studies and Toxicology, Commission on Life Sciences. Washington, DC: National Academy Press.

NRC (2009) Science and Decisions: Advancing Risk Assessments. Committee on Improving Risk Analysis Approaches Used by the U.S. EPA, Board on Environmental Studies and Toxicology, Division on Earth and Life Studies. Washington, DC: the National Academies Press.

US EPA. (2000) Science Policy Council Handbook: Risk Characterization. Washington, DC: Office of Research and Development. EPA 100-B-00-002.

US EPA Guidelines and Risk Assessment Forum Technical Reports (http://www.epa.gov/iris/backgrd.html).

INDEX

D

E

J

K

L

M

N

S

T

U

V

W